SOLIDWORKS® 2024
Quick Start

David C. Planchard
CSWP & SOLIDWORKS Accredited Educator

PUBLICATIONS

SDC Publications

P.O. Box 1334

Mission, KS 66222

913-262-2664

www.SDCpublications.com

Download all needed model files from the SDC Publication website (www.SDCpublications.com/downloads/978-1-63057-637-0).

ISBN-13: 978-1-63057-637-0

ISBN-10: 1-63057-637-9

Printed and bound in the United States of America.

INTRODUCTION

SOLIDWORKS® 2024 Quick Start is a great starting point for those new to SOLIDWORKS. Become familiar with the software interface and basic and advanced commands. Learn and apply proper design intent using various techniques to accomplish a particular task.

The book provides a step-by-step project-based learning approach.

Desired outcomes and usage competencies are listed for each chapter. The book is divided into three sections with 7 chapters and a bonus eBook on SOLIDWORKS and the **3DEXPERIENCE** platform.

Chapter 1 - 5: Explore the SOLIDWORKS User Interface, CommandManager and Document and System properties. Create simple and complex parts and assemblies. Learn and apply proper design intent. Create part configurations, multi-view drawings and a Bill of Materials (BOM) with Custom Properties.

Develop a mini–Stirling Engine and investigate the proper design intent and constraints. Explore view orientation based on the Front, Top and Right planes.

As you sketch, use tools such as the circle, line, centerline, slot and mirror. Modify sketches and sketch planes. Set document properties, identify sketch states and insert geometric relations and dimensions along with applying proper design technique.

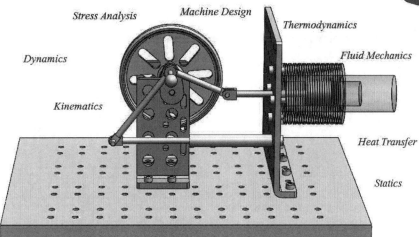

Parts consist of features. Features add or remove material. Apply and edit the following features: Extruded Boss/Base, Revolved Boss/Base, Extruded Cut, Circular Pattern, Hole Wizard and Fillet.

Explore and apply the Mass Properties, Measure and Appearance tool.

Assemblies consist of components and sub-assemblies. Incorporate a series of provided parts and your own parts to create two assemblies utilizing the Bottom-up approach.

Utilize the following assembly tools: Insert Component, Mate, Hide, Show, Rotate, Move, Modify, Flexible, Ridge and Multiple mate.

Learn how to add constraints that result in dynamic behavior of the assembly such as linear translation and rotation.

Before you machine or create a rapid prototype of a part for an assembly, verify clearance, interference, static and dynamic behavior between the assembly parts.

Apply the Assembly Visualization tool to sort components by mass while creating a motion study (animation) for a formal presentation.

Create two new drawings: Fly Wheel assembly and Bushing part. The first drawing is an Isometric Exploded view of the Fly Wheel assembly. The assembly drawing displays a Bill of Materials (BOM) at the part level along with Balloons, Magnetic lines and Custom Properties.

The second drawing is a Bushing part. Insert standard Orthographic (Front, Top & Right) views along with an Isometric view. Insert all needed dimensions, annotations and Custom Properties.

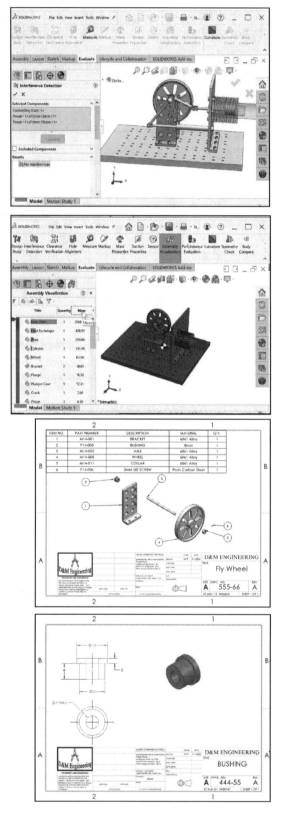

Chapter 6: Certified SOLIDWORKS Associate - Mechanical Design (CSWA) chapter provides detailed information and resources about the exam.

Details are provided on the login procedure; 3DEXPERIENCE® Certification Center at (https://3dexperience.virtualtester.com/#home) for instructors (if your school is an academic certification provider) to allocate free exam credits for Segment 1 & 2 of the CSWA exam.

Login information and procedure is provided for students to register, examine their certifications, manage their account and take the exam.

Material is provided for the free online student CSWA sample exam.

Note: The book provides the online CSWA sample exam in PDF format along with the SOLIDWORKS models, solutions and additional sample exam examples.

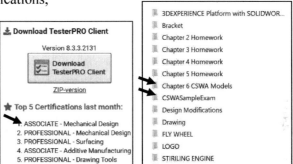

Chapter 7: Additive Manufacturing - 3D printing fundamentals chapter provides a basic understanding between Additive vs. Subtractive manufacturing. Discuss Fused Filament Fabrication (FFF), STereoLithography (SLA) and Selective Laser Sintering (SLS) printer technology.

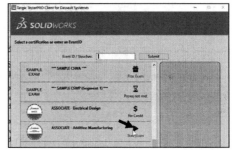

Select suitable filament material. Comprehend 3D printer terminology. Knowledge of preparing, saving and printing a model on a Fused Filament Fabrication 3D printer. Information on the Certified SOLIDWORKS Associate Additive Manufacturing (CSWA-AM) exam and general topic areas.

Details are provided on the login procedure; 3DEXPERIENCE® Certification Center at (https://3dexperience.virtualtester.com/#home) for instructors (if your school is an academic certification provider) to allocate free exam credits for the CSWA-AM exam.

Login information and procedure is provided for students to register, examine their certifications, manage their account and take the exam.

Bonus Section: SOLIDWORKS and the **3D**EXPERIENCE platform.

Detailed information is provided to obtain a **3D**EXPERIENCE ID (login credentials) to access the platform.

Install the Design with SOLIDWORKS App. This provides the ability to upload, store, manage, and share files.

The platform enables you to collaborate with students and teams on and off campus through a browser. Think of the **3D**EXPERIENCE platform simply as a SOLIDWORKS Add-In on your desktop.

Experience a single modeling environment. Access the platform within the SOLIDWORKS Task Pane.

Nine step-by-step tutorials and models are provided to practice and understand the tools and advantages of SOLIDWORKS and the platform.

About the Author

David Planchard is the founder of D&M Education LLC. Before starting D&M Education, he spent over 35 years in industry and academia holding various engineering, marketing, and teaching positions. He holds five U.S. patents. David has published and authored numerous papers and blogs on Machine Design, Product Design, Mechanics of Materials, Solid Modeling, SOLIDWORKS xDesign, SOLIDWORKS and the 3DEXPERIENCE Platform by Dassault Systèmes. David holds a BSME, MSM with the following professional certifications: CCAI, CCNP, CSWA-SD, CSWA-S, CSWA-AM, CSWP, CSWP-DRWT, SOLIDWORKS Accredited Educator, **3D**EXPERIENCE 3DSwymer, **3D**EXPERIENCE Collaborative Industry Innovator, **3D**EXPERIENCE Collaborative Designer for SOLIDWORKS and **3D**EXPERIENCE SOLIDWORKS Professional.

David is a SOLIDWORKS Solution Partner, **3D**EXPERIENCE Edu Champion and an Affiliate Professor of Emeritus in the Mechanical and Materials Engineering Department at WPI.

In 2012, David's senior Major Qualifying Project team (senior capstone) won first place in the Mechanical Engineering department at WPI.

In 2014, 2015 and 2016, David's senior Major Qualifying Project teams won the Provost award in the Mechanical Engineering department for design excellence.

In 2018, David's senior Major Qualifying Project team (Co-advisor) won the Provost award in the Electrical and Computer Engineering department. Subject area: Electrical System Implementation of Formula SAE Racing Platform.

In 2020, David was awarded Emeritus status at Worcester Polytechnic Institute (WPI) in the Department of Mechanical Engineering. David's ME design class achieved world recognition, featured in Compass published by Dassault Systèmes, in the technical article, "Open Innovation in a Pandemic."

In 2022, David's senior Major Qualifying Project team won second place in the Provost award in the Electrical and Computer Engineering department (ECE) at WPI. His FSAE Electric team won the IEEE Excellence in Electric Vehicle Engineering Award and achieved 4th place at the Formula Hybrid + Electric competition at the New Hampshire Motor Speedway in Loudon, NH. The competition is part of the Society of Automotive Engineers (SAE) Collegiate Design Series and is regarded as the most complex and dynamic of the series.

In 2023, David's senior Major Qualifying Project team won second place in the Provost award in the Electrical and Computer Engineering department (ECE) at WPI. His FSAE Electric team won the IEEE Excellence in Electric Vehicle Engineering Award, the TestEquity Electric Design award, and achieved 3rd place at the Formula Hybrid + Electric competition at the New Hampshire Motor Speedway in Loudon, NH.

In 2023, David was awarded the title of Affiliate Professor Emeritus in the Department of Mechanical and Materials Engineering at WPI.

David Planchard is the author of the following books:

- **Engineering Design with SOLIDWORKS® 2024**, 2023, 2022, 2021, 2020, 2019, 2018 and 2017

- **Engineering Graphics with SOLIDWORKS® 2024**, 2023, 2022, 2021, 2020, 2019, 2018 and 2017

- **SOLIDWORKS® 2024 Quick Start**, 2023, 2022, 2021, 2020, 2019 and 2018

- **SOLIDWORKS® 2024 Tutorial**, 2023, 2022, 2021, 2020, 2019 and 2017

- **Drawing and Detailing with SOLIDWORKS® 2022**, 2014 and 2012

- **Official Certified SOLIDWORKS® Professional (CSWP) Certification Guide 2020 - 2023**, 2019 - 2020 and 2015 - 2017

- **Official Guide to Certified SOLIDWORKS® Associate Exams: CSWA, CSWA-SD, CSWA-S, CSWA-AM 2020 - 2023**, 2019 - 2021, 2017 - 2019 and 2015 - 2017

Acknowledgements

Writing this book was a substantial effort that would not have been possible without the help and support of my loving family and of my professional colleagues. I would like to thank Professor John M. Sullivan Jr., Professor Jack Hall, and the community of scholars at Worcester Polytechnic Institute who have enhanced my life, my knowledge and helped to shape the approach and content to this text.

The author is greatly indebted to my colleagues from Dassault Systèmes SOLIDWORKS Corporation for their help and continuous support: Mike Puckett, Avelino Rochino, Yannick Chaigneau, Terry McCabe and the SOLIDWORKS Partner team.

Thanks also to Professor Richard L. Roberts of Wentworth Institute of Technology, Professor Dennis Hance of Wright State University, Professor Jason Durfess of Eastern Washington University and Professor Aaron Schellenberg of Brigham Young University - Idaho who provided vision and invaluable suggestions.

Contact the Author

We realize that keeping software application books current is imperative to our customers. We value the hundreds of professors, students, designers and engineers that have provided us input to enhance the book. Please contact me directly with any comments, questions or suggestions on this book or any of our other SOLIDWORKS books at dplanchard@msn.com.

Note to Instructors

Download all needed model files (initial and final) from the SDC Publications website (www.SDCpublications.com/downloads/978-1-63057-637-0).

Instructors are provided PowerPoint presentations and chapter quizzes. Initial and final SOLIDWORKS models are included for each chapter and homework section.

Over 100 additional models are provided as classroom examples using various SOLIDWORKS features, sketches and mate types. Additional design projects are included.

The sample online CSWA exam is provided in the book (PDF) format along with the SOLIDWORKS models, solutions and additional practice exam problems.

An instructor's folder is provided with log in and allocating student exam credit information.

An eBook for SOLIDWORKS and the **3D**EXPERIENCE platform is included.

Trademarks, Disclaimer and Copyrighted Material

© Dassault Systèmes. All rights reserved. **3D**EXPERIENCE, the 3DS logo, the Compass icon, IFWE, 3DEXCITE, 3DVIA, BIOVIA, CATIA, CENTRIC PLM, DELMIA, ENOVIA, GEOVIA, MEDIDATA, NETVIBES, OUTSCALE, SIMULIA and SOLIDWORKS are commercial trademarks or registered trademarks of Dassault Systèmes, a European company (Societas Europaea) incorporated under French law, and registered with the Versailles trade and companies registry under number 322 306 440, or its subsidiaries in the United States and/or other countries.

All statements are strictly based on the author's opinion. Dassault Systèmes and its affiliates disclaim any liability, loss, or risk incurred as a result of the use of any information or advice contained in this book, either directly or indirectly.

SOLIDWORKS®, eDrawings®, SOLIDWORKS Simulation®, SOLIDWORKS Flow Simulation, and SOLIDWORKS Sustainability are a registered trademark of Dassault Systèmes SOLIDWORKS Corporation in the United States and other countries; certain images of the models in this publication courtesy of Dassault Systèmes SOLIDWORKS Corporation.

The publisher and the author make no representations or warranties with respect to the accuracy or completeness of the contents of this work and specifically disclaim all warranties, including without limitation warranties of fitness for a particular purpose. No warranty may be created or extended by sales or promotional materials. Dimensions of parts are modified for illustration purposes. Every effort is made to provide an accurate text.

The authors and the manufacturers shall not be held liable for any parts, components, assemblies or drawings developed or designed with this book or any responsibility for inaccuracies that appear in the book. Web and company information was valid at the time of this printing.

References

- SOLIDWORKS Help Topics and What's New, SOLIDWORKS Corporation, 2024.
- 80/20 Product Manual, 80/20, Inc., Columbia City, IN, 2012.
- Ticona Designing with Plastics - The Fundamentals, Summit, NJ, 2009.
- Emerson-EPT Bearing Product Manuals and Gear Product Manuals, Emerson Power Transmission Corporation, Ithaca, NY, 2009.
- Emhart - A Black and Decker Company, On-line catalog, Hartford, CT, 2012.

During the initial SOLIDWORKS installation, you are requested to select either the ISO or ANSI drafting standard. ISO is typically a European drafting standard and uses First Angle Projection. The book is written using the ANSI (US) overall drafting standard and Third Angle Projection for drawings.

TABLE OF CONTENTS

Exclusive Bonus Content – Instructions for download on inside front cover of book

Overview of Chapters

Chapter 1: Overview of SOLIDWORKS and the User Interface

SOLIDWORKS is a design software application used to create 2D and 3D sketches, 3D parts and assemblies and 2D drawings.

Chapter 1 introduces the user to the SOLIDWORKS User Interface, CommandManager, Document and System properties. Review and address the: Menu bar toolbar, Menu bar menu, Drop-down menus, Context toolbars, Consolidated drop-down toolbars, System feedback icons, Confirmation Corner, Heads-up View toolbar and Document Properties.

Start a new SOLIDWORKS session. Create a new part. Open an existing part and view the created features and sketches using the Rollback bar. Review the design intent in the part.

Chapter 2: 2D Sketching, Features and Parts

Learn about 2D Sketching and 3D features. Create a new part called Wheel with user defined document properties.

Create the Wheel for the Fly Wheel sub-assembly. Utilize the Fly Wheel sub-assembly in the final Stirling Engine assembly.

Apply the following sketch and feature tools: Circle, Line Centerline, Centerpoint Straight Slot, Mirror Entities, Extruded Boss, Extruded Cut, Revolved Boss, Circular Pattern, Hole Wizard and Fillet.

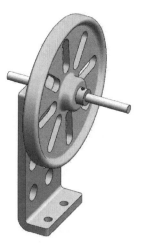

Incorporate design change into a part using proper design intent, along with applying multiple geometric relations: Coincident, Vertical, Horizontal, Tangent and Midpoint and feature and sketch modifications.

Utilize the Material, Mass Properties and Appearance tool on the Wheel.

Chapter 3: Assembly Modeling - Bottom-up method

Learn about the Bottom-up assembly method and create two new assemblies with user defined document properties:

- Fly Wheel.

- Stirling Engine.

Insert the following Standard mate types: Coincident, Concentric, Distance and Tangent.

Utilize the following assembly tools: Insert Component, Suppress, Un-suppress, Mate, Move Component, Rotate Component, Interference Detection, Hide, Show, Flexible, Ridge and Multiple mate.

Create an Exploded View with animation.

Apply the Measure and Mass Properties tool to modify a component in the Stirling Engine assembly.

Chapter 4: Design Modifications

Address clearance, interference, static and dynamic behavior of the Stirling Engine Modified assembly.

Verify the behavior between the following components: Power Piston, Power Clevis, Connecting Rod and Handle in the assembly.

Apply the following assembly tools: Move, Rotate, Collision Detection, Interference Detection, Selected Components, Edit Mate and Center of Mass.

Utilize the Assembly Visualization tool on the assembly and sort by component mass.

Create a new Coordinate System relative to the default origin.

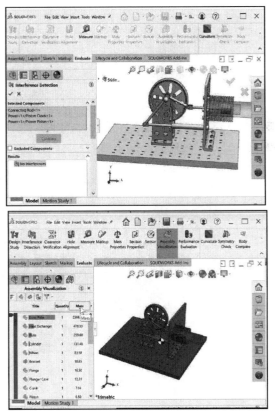

Additional projects are included in the exercise section of chapter 4. Use the web to locate a similar assembly and to understand the parts needed in the assembly. View the provided sample parts. Create your own parts and final assembly. View the provided motion file to create the proper movement in SOLIDWORKS.

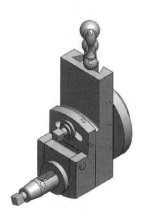

- Arbor Press Project
- Bench Vice Project
- Butterfly Valve Project
- Drill Guide Project
- Kant Twist Clamp Project
- Pipe Vice Project
- Pulley Project
- Quick Acting Clamp Project
- Radial Engine Project
- Shaper Tool Head Project
- Welder Arm Project

Chapter 5: Drawing and Dimensioning Fundamentals

Learn about Drawing and Dimension Fundamentals and create two new drawings with user defined document properties:

- Fly Wheel Assembly.

- Bushing.

Create the Fly Wheel Assembly drawing with an Exploded Isometric view.

Utilize a Bill of Materials (BOM), Magnetic lines and balloons.

Learn and apply Custom Properties and the Title Block.

Create the Bushing Part drawing utilizing Third Angle Projection with three views: Front, Top and Isometric.

Address imported dimensions from the Model Items tool.

Insert additional dimensions using the Smart Dimension tool along with all needed annotations.

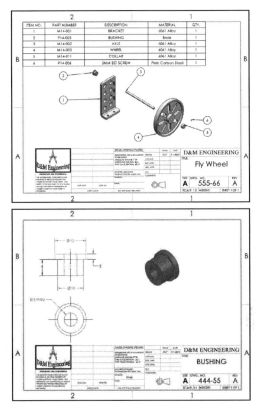

Chapter 6: Certified SOLIDWORKS Associate - Mechanical Design (CSWA) exam

Chapter 6 provides general information and resources for instructors and students about the Certified SOLIDWORKS Associate - Mechanical Design (CSWA) exam.

Details are provided on the login procedure; 3DEXPERIENCE® Certification Center at (https://3dexperience.virtualtester.com/#home) for instructors (if your school is an academic certification provider) to allocate free exam credits for Segment 1 & 2 of the CSWA exam.

Login information and procedure is provided for students to register, examine their certifications, manage their account and take the exam.

Information is provided for the free online student CSWA sample exam. Note: The book provides the free online CSWA sample exam in pdf format along with the SOLIDWORKS models, solutions and additional sample exam examples.

Additional reference to help you prepare:

- **Official Guide to Certified SOLIDWORKS Associate Exams: CSWA, CSWA-SD, CSWA-S, CSWA-AM; 2020 -2023,** 2017-2019, 2015 - 2012. There are over 65 practice questions and examples (initial and final models) in the Official GUIDE to Certified SOLIDWORKS Exam book for the CSWA certification.

Download all needed model files from the SDC Publication website www.SDCpublications.com/downloads/978-1-63057-637-0.

Chapter 7: Additive Manufacturing - 3D Printing

Chapter 7 provides a basic understanding between Additive vs. Subtractive manufacturing. Discuss Fused Filament Fabrication (FFF), STereoLithography (SLA) and Selective Laser Sintering (SLS) printer technology.

Select suitable filament material. Comprehend 3D printer terminology. Knowledge of preparing, saving and printing a model on a Fused Filament Fabrication 3D printer.

Information on the Certified SOLIDWORKS Associate Additive Manufacturing (CSWA-AM) exam.

On the completion of the project, you will be able to:

- Discuss Additive vs Subtractive manufacturing.

- Review 3D printer technology: Fused Filament Fabrication (FFF), STereoLithography (SLA) and Selective Laser Sintering (SLS).

- Select the correct filament material: PLA (Polylactic acid), FPLA (Flexible Polylactic acid), ABS (Acrylonitrile butadiene styrene), PVA (Polyvinyl alcohol), Nylon 618 and Nylon 645.

- Create an STL (*.stl) file, Additive Manufacturing (*.amf) file and a 3D Manufacturing format (*.3mf) file.

- Prepare Geometric Code, "G-Code".

- Comprehend general 3D printer terminology.

- Understand optimum build orientation.

- Enter slicer parameters:

  o Raft, brim, skirt, layer height, percent infill, infill pattern, wall thickness, fan speed, print speed, bed temperature and extruder (hot end) temperature.

- Address tolerance for interlocking parts.

- Knowledge of the Certified SOLIDWORKS Associate Additive Manufacturing exam.

If your school is an academic certification provider, your instructor can allocate free exam credits for the CSWA (Segment 1 & 2), CSWA-SD, CSWA-S, and CSWA-AM certifications. The instructor will require your .edu email address.

Login information and procedure is provided for students to register, examine their certifications, manage their account and take the exam.

Book Layout

The following conventions are used throughout this book:

- The term document is used to refer to a SOLIDWORKS part, drawing or assembly file.

- The list of items across the top of the SOLIDWORKS interface is the Menu bar menu or the Menu bar toolbar. Each item in the Menu bar has a pull-down menu. When you need to select a series of commands from these menus, the following format is used: Click **View**, **Hide/Show**, check **Origins** from the Menu bar. The Origins are displayed in the Graphics window.

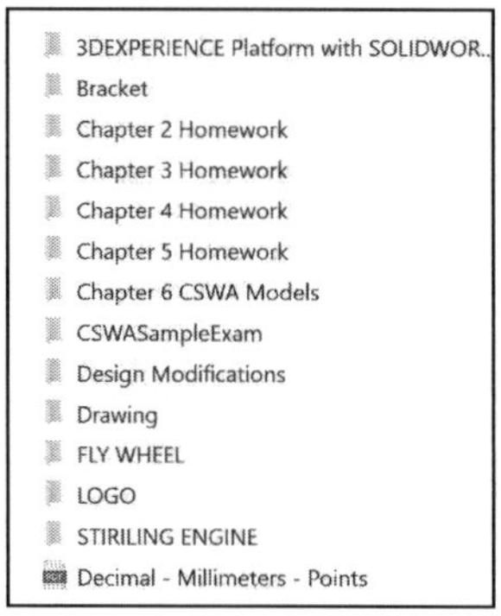

- The ANSI overall drafting standard and Third Angle projection is used as the default setting in this text. MMGS (millimeter, gram, second) unit system is used.

3DEXPERIENCE Platform with SOLIDWOR...
Bracket
Chapter 2 Homework
Chapter 3 Homework
Chapter 4 Homework
Chapter 5 Homework
Chapter 6 CSWA Models
CSWASampleExam
Design Modifications
Drawing
FLY WHEEL
LOGO
STIRILING ENGINE
Decimal - Millimeters - Points

- The book is organized into various chapters. Each chapter is focused on a specific subject or feature.

- All templates, logos and needed model documents for this book are provided.

- Download model files from the SDC Publications website (www.SDCpublications.com/downloads/978-1-63057-637-0.)

- The platform refers to the **3D**EXPERIENCE platform that provides downloadable content (DLC) from a secure cloud location within your SOLIDWORKS session.

- The following command syntax is used throughout the text. Commands that require you to perform an action are displayed in **Bold** text.

Format:	Convention:	Example:
Bold	- All commands actions. - Selected icon button. - Selected geometry: line, circle. - Value entries.	- Click **Options** ⚙ ▾ from the Menu bar toolbar. - Click **Corner Rectangle** ▾ from the Sketch toolbar. - Click **Sketch** from the Context toolbar. - Select the **centerpoint**. - Enter **3.0** for Radius.
Capitalized	- Filenames. - First letter in a feature name.	- Save the **FLATBAR** assembly. - Click the **Fillet** feature.

Windows Terminology in SOLIDWORKS

The mouse buttons provide an integral role in executing SOLIDWORKS commands. The mouse buttons execute commands, select geometry, display Shortcut menus and provide information feedback.

A summary of mouse button terminology is displayed below:

Item:	Description:
Click	Press and release the left mouse button.
Double-click	Double press and release the left mouse button.
Click inside	Press the left mouse button. Wait a second, and then press the left mouse button inside the text box. Use this technique to modify Feature names in the FeatureManager design tree.
Drag	Point to an object, press and hold the left mouse button down. Move the mouse pointer to a new location. Release the left mouse button.
Right-click	Press and release the right mouse button. A Shortcut menu is displayed. Use the left mouse button to select a menu command.
Tool Tip	Position the mouse pointer over an Icon (button). The tool name is displayed below the mouse pointer.
Large Tool Tip	Position the mouse pointer over an Icon (button). The tool name and a description of its functionality are displayed below the mouse pointer.
Mouse pointer feedback	Position the mouse pointer over various areas of the sketch, part, assembly or drawing. The cursor provides feedback depending on the geometry.

A mouse with a center wheel provides additional functionality in SOLIDWORKS. Roll the center wheel downward to enlarge the model in the Graphics window. Hold the center wheel down. Drag the mouse in the Graphics window to rotate the model.

Visit SOLIDWORKS website:
http://www.SOLIDWORKS.com/sw/support/hardware.html to view their supported operating systems and hardware requirements.

The book is designed to expose the new user to numerous tools and procedures. It may not always use the simplest and most direct process.

Notes:

Chapter 1

Overview of SOLIDWORKS® 2024 and the User Interface

Below are the desired outcomes and usage competencies based on the completion of Chapter 1.

Desired Outcomes:	Usage Competencies:
• A comprehensive understanding of the SOLIDWORKS® User Interface (UI) and CommandManager.	• Ability to establish a SOLIDWORKS session. • Aptitude to utilize the following items: Menu bar toolbar, Menu bar menu, Drop-down menus, Context toolbars, Consolidated drop-down toolbars, System feedback icons, Task Pane, Confirmation Corner, Heads-up View toolbar and System and Document Properties. • Open a new and existing SOLIDWORKS part. • Knowledge to zoom, rotate and maneuver a three-button mouse in the SOLIDWORKS Graphics window.

Notes:

Chapter 1 - Overview of SOLIDWORKS® 2024 and the User Interface

Chapter Objective

Chapter 1 introduces the user to the SOLIDWORKS User Interface, CommandManager, Document and System properties. Review and address: Menu bar toolbar, Menu bar menu, Drop-down menus, Context toolbars, Consolidated drop-down toolbars, System feedback icons, Confirmation Corner and Heads-up View toolbar.

Start a new SOLIDWORKS session. Open an existing part. View the created features and sketches using the Rollback bar. Examine design intent.

On the completion of this chapter, you will be able to:

- Utilize the Welcome - SOLIDWORKS dialog box.

- Establish a SOLIDWORKS session.

- Comprehend the SOLIDWORKS User Interface.

- Recognize the default Reference Planes in the FeatureManager.

- Open a new and existing SOLIDWORKS part.

- Utilize Help and SOLIDWORKS Tutorials.

- Zoom, rotate and maneuver a three-button mouse in the SOLIDWORKS Graphics window.

What is SOLIDWORKS®?

- SOLIDWORKS® is a design tool by Dassault Systèmes. The **3D**EXPERIENCE® platform is a technology platform that connects SOLIDWORKS to other design, simulation, intelligent information and collaboration Apps through a web browser.

- SOLIDWORKS® is a mechanical design automation software package used to build parts, assemblies and drawings that takes advantage of the familiar Microsoft® Windows graphical user interface.

- SOLIDWORKS is an easy to learn design and analysis tool (SOLIDWORKS Simulation, SOLIDWORKS Motion, SOLIDWORKS Flow Simulation, Sustainability, etc.), which makes it possible for designers to quickly sketch 2D and 3D concepts, create 3D parts and assemblies and detail 2D drawings.

- Model dimensions in SOLIDWORKS are associative between parts, assemblies and drawings. Reference dimensions are one-way associative from the part to the drawing or from the part to the assembly.

- This book is written for the beginner to intermediate user.

Start a SOLIDWORKS Session

Start a SOLIDWORKS session and familiarize yourself with the SOLIDWORKS User Interface. As you read and perform the tasks in this chapter, you will obtain a sense of how to use the book and the structure. Actual input commands or required actions in the chapter are displayed in bold.

The book does not cover starting a SOLIDWORKS session in detail for the first time. A default SOLIDWORKS installation presents you with several options. For additional information, visit http://www.SOLIDWORKS.com.

Activity: Start a SOLIDWORKS Session.

Start a SOLIDWORKS session.

1) Type **SOLIDWORKS 2024** in the Search window.

2) Click the **SOLIDWORKS 2024** application (or if available, **double-click** the SOLIDWORKS icon on the desktop). The Welcome - SOLIDWORKS dialog box is displayed by default.

The Welcome - SOLIDWORKS box provides a convenient way to open recent documents (Parts, Assemblies and Drawings), view recent folders, access SOLIDWORKS resources, and stay updated on SOLIDWORKS news.

3) **View** your options. Do not open a document at this time. **Note**: Your Task Pane icon order may be different..

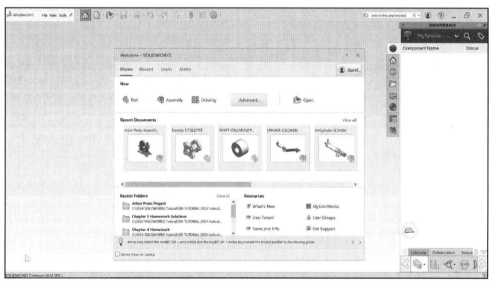

Home Tab

The Home tab lets you open new and existing documents, view recent documents and folders, and access SOLIDWORKS resources (*Part, Assembly, Drawing, Advanced mode, Open*).

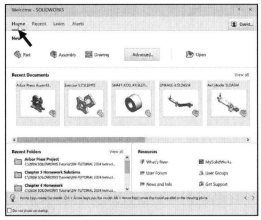

Recent Tab

The Recent tab lets you view a longer list of recent documents and folders. Sections in the Recent tab include *Documents* and *Folders*.

The Documents section includes thumbnails of documents that you have opened recently.

Click a thumbnail to open the document, or hover over a thumbnail to see the document location and access additional information about the document. When you hover over a thumbnail, the full path and last saved date of the document appears.

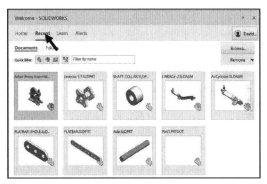

Learn Tab

The Learn tab lets you access instructional resources to help you learn more about the SOLIDWORKS software.

Sections in the Learn tab include:

- **Get Started with SOLIDWORKS CAD**.

- **Get Started with Collaborative Designer for SOLIDWORKS (Learning Path)**.

- **MySolidWorks Training (Learning Portal)**.

- **MySolidWorks CAD models**.

- **On my PC**.

- **Getting Started guide**.

- **My Training**. Open the My Training section at MySolidWorks.com.

- **Certification**. Open the SOLIDWORKS Certification Program section at solidworks.com. You will need to create an account with a password.

- **Curriculum**. Open the Curriculum section at solidworks.com. You will need to create an account with a password.

💡 When you install the software, if you do not install the Help Files or Example Files, the Tutorials and Samples links are unavailable.

Alerts Tab

The Alerts tab keeps you updated with SOLIDWORKS news.

Sections in the Alerts tab include Critical, Troubleshooting, and Technical.

💡 The Critical section does not appear if there are no critical alerts to display.

- **Troubleshooting**. Includes troubleshooting messages and recovered documents that used to be on the SOLIDWORKS Recovery tab in the Task Pane.

💡 If the software has a technical problem and an associated troubleshooting message exists, the Welcome dialog box opens to the Troubleshooting section automatically on startup, even if you selected **Do not show at startup** in the dialog box.

- **Technical Alerts**. Open the contents of the SOLIDWORKS Support Bulletins RSS feed (Hotfixes, release news) at solidworks.com.

Close the Welcome - SOLIDWORKS dialog box.

4) Click **Close** ✕ from the Welcome - SOLIDWORKS dialog box. The SOLIDWORKS Graphics window is displayed. **Note**: You can also click outside the Welcome - SOLIDWORKS dialog box, in the Graphics window.

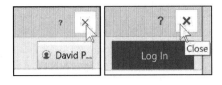

View the SOLIDWORKS Graphics window. **Note**: Your Task Pane icon order may be different.

Menu Bar toolbar

The SOLIDWORKS (UI) is designed to make maximum use of the Graphics window. The Menu Bar toolbar contains a set of the most frequently used tool buttons from the Standard toolbar.

The following default tools are available:

- **Welcome to SOLIDWORKS** ⌂ - Open the Welcome dialog box; **New** ☐ - Create a new document; **Open** 📂 - Open an existing document; **Save** 💾 - Save an active document; **Print** 🖨 - Print an active document; **Undo** ↺ - Reverse the last action; **Redo** ↻ - Redoes the last action that you reverse; **Select** ⬚ - Select Sketch entities, components and more; **Rebuild** ● - Rebuild the active part, assembly or drawing;

File Properties ▦ - Summary information on the active document; **Options** ⚙▾ - Change system options and Add-Ins for SOLIDWORKS; **Login** ⊚ - Login to SOLIDWORKS. You will need to create an account with a password; **Help** ⑦ - access to help, tutorials, updates and more.

Menu Bar menu (No model open)

SOLIDWORKS provides a context-sensitive menu structure. The menu titles remain the same for all three types of documents (Parts, Assemblies and Drawings), but the menu items change depending on which type of document is active.

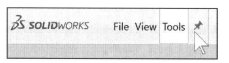

Menu Bar menu (Model open)

The Pin ⚲ option displays the Menu bar toolbar and the Menu bar menu as illustrated. Throughout the book, the Menu bar menu and the Menu bar toolbar are referred to as the Menu bar.

Drop-down menu (Open part document)

SOLIDWORKS takes advantage of the familiar Microsoft® Windows user interface. Communicate with SOLIDWORKS through drop-down menus, Context sensitive toolbars, Consolidated toolbars or the CommandManager tabs.

To close a SOLIDWORKS drop-down menu, press the Esc key. You can also click any other part of the SOLIDWORKS Graphics window or click another drop-down menu.

Create a New Part Document

Activity: Create a new Part Document.

A part is a 3D model, which consists of features. What are features? Features are geometry building blocks. Most features either add or remove material. Some features do not affect material (Cosmetic Thread).

Features are created either from 2D or 3D sketched profiles or from edges and faces of existing geometry.

Features are individual shapes that combined with other features make up a part or assembly. Some features, such as bosses and cuts, originate as sketches. Other features, such as shells and fillets, modify a feature's geometry.

The first sketch of a part is called the Base Sketch. The Base sketch is the foundation for the 3D model. The book focuses on 2D sketches and 3D features.

FeatureManager tabs, CommandManager tabs, Task Pane tabs and other tabs will vary depending on system setup, version and Add-ins.

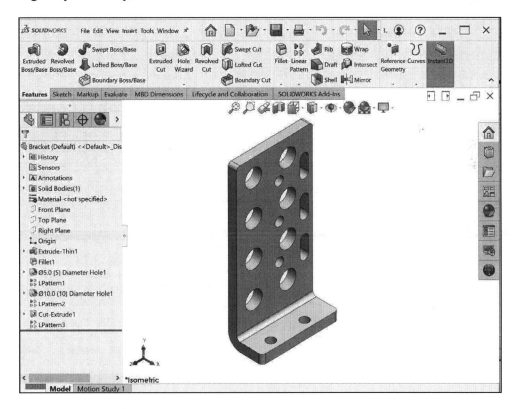

There are two modes in the New SOLIDWORKS Document dialog box: Novice and Advanced. The Novice option is the default option with three templates. The Advanced mode contains access to additional templates and tabs that you create in system options. Use the Advanced mode in this book.

Create a new part.

5) Click **New** 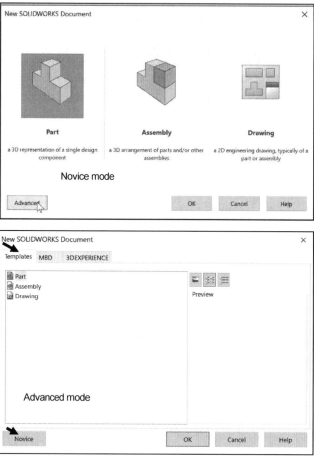 from the Menu bar. The New SOLIDWORKS Document dialog box is displayed.

Select the Advanced mode.

6) If needed, click the **Advanced** tab. The below New SOLIDWORKS Document box is displayed.

7) Click the **Templates** tab.

8) Click **Part**. Part is the default template from the New SOLIDWORKS Document dialog box.

9) Click **OK** from the New SOLIDWORKS Document dialog box.

The Advanced mode remains selected for all new documents in the current SOLIDWORKS session. When you exit SOLIDWORKS, the Advanced mode setting is saved.

The default SOLIDWORKS installation contains two tabs in the New SOLIDWORKS Document dialog box: Templates and MBD. The Templates tab corresponds to the default SOLIDWORKS templates. The MBD tab corresponds to the templates utilized in the SOLIDWORKS (Model Based Definition). You may see a Tutorial tab depending on your version.

Note: **3D**EXPERIENCE users can create templates for the **3D**EXPERIENCE platform directly from SOLIDWORKS. Use the 3DEXPERIENCE tab in the New SOLIDWORKS Document dialog box.

Part1 is displayed in the FeatureManager and is the name of the document. Part1 is the default part window name.

The Part Origin ⌐ is displayed in blue in the center of the Graphics window. The Origin represents the intersection of the three default reference planes: *Front Plane*, *Top Plane* and *Right Plane*. The positive X-axis is horizontal and points to the right of the Origin in the Front view. The positive Y-axis is vertical and points upward in the Front view. The FeatureManager contains a list of features, reference geometry, and settings utilized in the part.

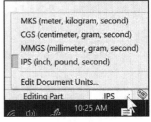

Edit the document units directly from the Graphics window as illustrated.

CommandManager tabs, FeatureManager tabs, Task Pane tabs will vary depending on system setup, version and Add-ins.

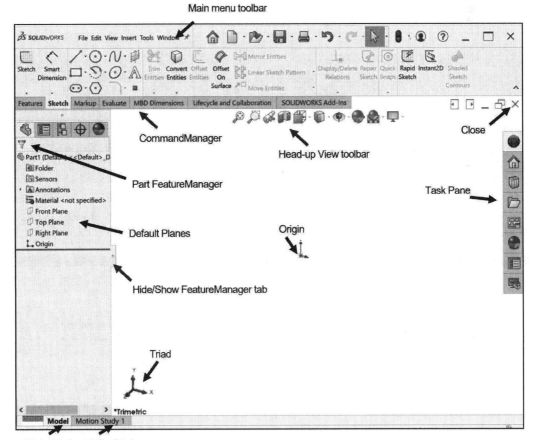

View the Default Sketch Planes.

10) Click the **Front Plane** from the FeatureManager.

11) Click the **Top Plane** from the FeatureManager.

12) Click the **Right Plane** from the FeatureManager.

13) Click the **Origin** from the FeatureManager. The Origin is the intersection of the Front, Top and Right Planes. The Origin point is displayed.

14) Click **inside** the Graphics window.

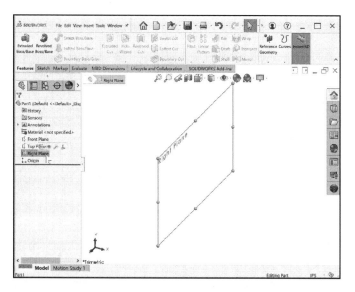

Download the SOLIDWORKS-MODELS 2024 folder from the SDC Publication website (www.SDCpublications.com/downloads/978-1-63057-637-0). Open the Bracket part. Review the features and sketches in the Bracket FeatureManager. Work directly from a local hard drive.

Activity: Download the SOLIDWORKS folder. Open the Bracket Part.

Download the SOLIDWORKS folder. Open an existing SOLIDWORKS part.

15) **Download** the SOLIDWORKS-MODELS 2024 folder.

16) **Unzip** the folder. **Work** from the unzip folder.

17) Click **Open** 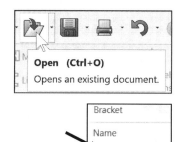 from the Menu bar menu.

18) Browse to the **SOLIDWORKS-MODELS-2024\Bracket** folder.

19) Double-click the **Bracket** part. The Bracket part is displayed in the Graphics window.

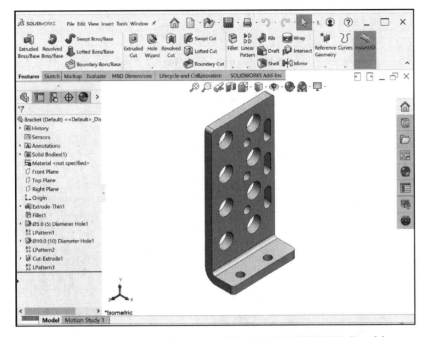

The FeatureManager design tree is located on the left side of the SOLIDWORKS Graphics window. The FeatureManager provides a summarized view of the active part, assembly, or drawing document. The tree displays the details on how the part, assembly or drawing document was created.

Use the FeatureManager rollback bar to temporarily roll back to an earlier state, to absorbed features, roll forward, roll to previous, or roll to the end of the FeatureManager design tree. You can add new features or edit existing features while the model is in the rolled-back state. You can save models with the rollback bar placed anywhere.

In the next section, review the features in the Bracket FeatureManager using the Rollback bar.

Activity: Use the FeatureManager Rollback Bar option.

Apply the FeatureManager Rollback Bar. Revert to an earlier state in the model.

20) Place the **mouse pointer** over the rollback bar in the FeatureManager design tree as illustrated. The pointer changes to a hand 🖐. Note the provided information on the feature. This is called Dynamic Reference Visualization.

21) Drag the **rollback bar** up the FeatureManager design tree until it is above the features you want rolled back, in this case 10.0 (10) Diameter Hole1.

22) **Release** the mouse button.

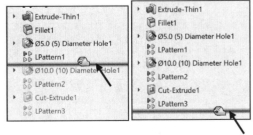

View the first feature in the Bracket Part.

23) Drag the **rollback bar** up the FeatureManager above Fillet1. View the results in the Graphics window. This is a great feature to re-engineer a part.

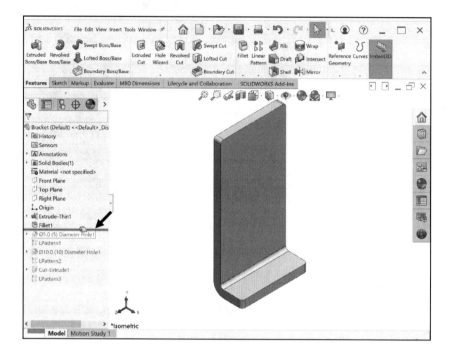

Return to the original Bracket Part FeatureManager.

24) Right-click **Extrude-Thin1** in the FeatureManager. The Pop-up Context toolbar is displayed.

25) Click **Roll to End**. View the results in the Graphics window.

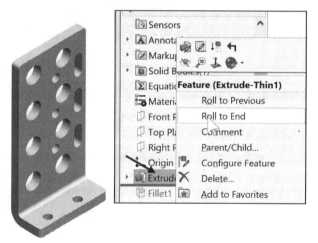

Heads-up View toolbar

SOLIDWORKS provides the user with numerous view options. One of the most useful tools is the Heads-up View toolbar displayed in the Graphics window when a document is active.

💡 *Dynamic Annotation Views* 🔍: Only available with SOLIDWORKS MBD (Model Based Definition). Provides the ability to control how annotations are displayed when you rotate models.

In the next section, apply the following tools: Zoom to Fit, Zoom to Area, Zoom out, Rotate and select various view orientations from the Heads-up View toolbar.

> **Activity: Utilize the Heads-up View toolbar**.

Zoom to Fit the model in the Graphics window.

26)   Click the **Zoom to Fit** 🔍 icon. The tool fits the model to the Graphics window.

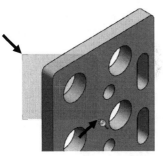

Zoom to Area on the model in the Graphics window.

27)   Click the **Zoom to Area** 🔍 icon. The Zoom to Area 🔍 icon is displayed.

Zoom in on the top left hole.

28)   **Window-select** the top left corner as illustrated. View the results.

De-select the Zoom to Area tool.

29)   Click the **Zoom to Area** 🔍 icon.

Fit the model to the Graphics window.

30)   Press the **f** key.

Rotate the model.

31)   Hold the **middle mouse button** down. Drag **upward** ↻, **downward** ↻, to the **left** ↻ and to the **right** ↻ to rotate the model in the Graphics window.

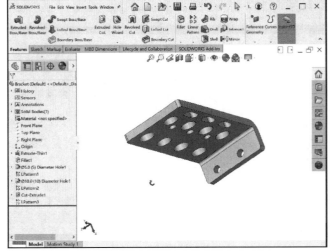

Display a few Standard Views.

32) Click **inside** the Graphics window.

33) Click **Front** ⬜ from the drop-down Heads-up view toolbar. The model is displayed in the Front view.

34) Click **Right** ⬜ from the drop-down Heads-up view toolbar. The model is displayed in the Right view.

35) Click **Top** ⬜ from the drop-down Heads-up view toolbar. The model is displayed in the Top view.

Display a Trimetric view of the Bracket model.

36) Click **Trimetric** ⬜ from the drop-down Heads-up view toolbar as illustrated. Note your options. View the results in the Graphics window.

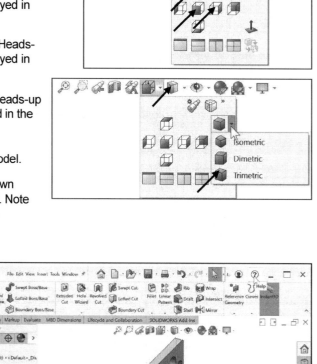

SOLIDWORKS Help

Help in SOLIDWORKS is context-sensitive and in HTML format. Help is accessed in many ways, including Help buttons in all dialog boxes and PropertyManager and Help ⑦ on the Standard toolbar for SOLIDWORKS Help.

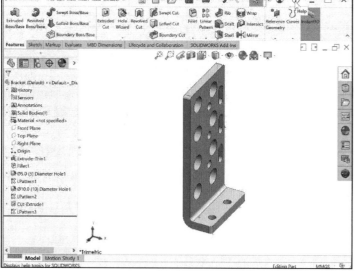

37)   Click ⑦ from the Standard toolbar.

38)   Click ⑦ **Help** from the drop-down menu. The SOLIDWORKS Home Page is displayed by default. View your options.

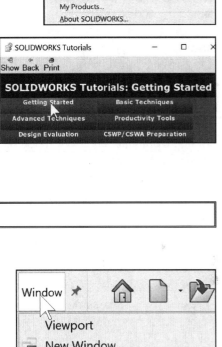

💡 SOLIDWORKS Web Help is active by default under Help in the Main menu.

Close Help. Return to the SOLIDWORKS Graphics window.

39)   **Close** ❌ SOLIDWORKS Home.

SOLIDWORKS Tutorials

Display and explore the SOLIDWORKS tutorials.

40)   Click ⑦ from the Standard toolbar.

41)   Click **Tutorials**. The SOLIDWORKS Tutorials are displayed. The SOLIDWORKS Tutorials are presented by category.

42)   Click the **Getting Started** category. The Getting Started category provides lessons on parts, assemblies, and drawings.

In the next section, close all models, tutorials and view the additional User Interface tools.

> **Activity: Close all Tutorials and Models**.

Close SOLIDWORKS Tutorials and models.

43)   **Close** ❌ SOLIDWORKS Tutorials.

44)   Click **Window**, **Close All** from the Menu bar menu.

User Interface Tools

The book utilizes additional areas of the SOLIDWORKS User Interface. Explore an overview of these tools in the next section.

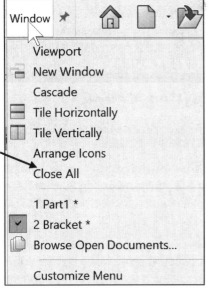

Right-click

Right-click in the Graphics window on a model, or in the FeatureManager on a feature or sketch to display the Context-sensitive toolbar. If you are in the middle of a command, this toolbar displays a list of options specifically related to that command.

Right-click an empty space in the Graphics window of a part or assembly, and a selection context toolbar above the shortcut menu is displayed. This provides easy access to the most commonly used selection tools.

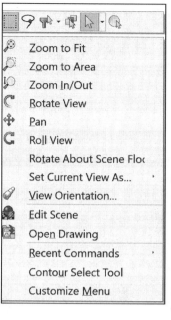

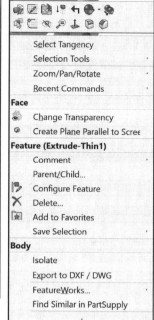

Consolidated toolbar

Similar commands are grouped together in the CommandManager. For example, variations of the Rectangle sketch tool are grouped in a single fly-out button as illustrated.

If you select the Consolidated toolbar button without expanding:

For some commands such as Sketch, the most commonly used command is performed. This command is the first listed and the command shown on the button.

For commands such as rectangle, where you may want to repeatedly create the same variant of the rectangle, the last used command is performed. This is the highlighted command when the Consolidated toolbar is expanded.

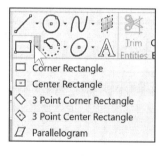

System feedback icon

SOLIDWORKS provides system feedback by attaching a symbol to the mouse pointer cursor.

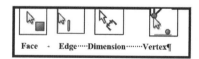

The system feedback symbol indicates what you are selecting or what the system is expecting you to select.

As you move the mouse pointer across your model, system feedback is displayed in the form of a symbol, riding next to the cursor as illustrated. This is a valuable feature in SOLIDWORKS.

Confirmation Corner

When numerous SOLIDWORKS commands are active, a symbol or a set of symbols is displayed in the upper right-hand corner of the Graphics window. This area is called the Confirmation Corner.

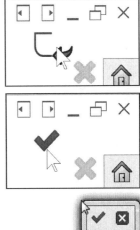

When a sketch is active, the confirmation corner box displays two symbols. The first symbol is the sketch tool icon. The second symbol is a large red X. These two symbols supply a visual reminder that you are in an active sketch. Click the sketch symbol icon to exit the sketch and to save any changes that you made.

When other commands are active, the confirmation corner box provides a green check mark and a large red X. Use the green check mark to execute the current command. Use the large red X to cancel the command.

Confirm changes you make in sketches and tools by using the D keyboard shortcut to move the OK and Cancel buttons to the pointer location in the Graphics window.

Heads-up View toolbar

SOLIDWORKS provides the user with numerous view options from the Standard Views, View and Heads-up View toolbar.

The Heads-up View toolbar is a transparent toolbar that is displayed in the Graphics window when a document is active.

You can hide, move or modify the Heads-up View toolbar. To modify the Heads-up View toolbar, right-click on a tool and select or deselect the tools that you want to display.

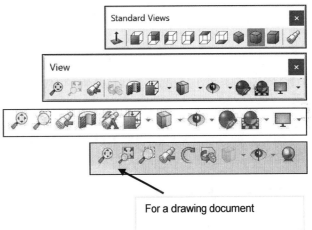

For a drawing document

The following views are available.
Note: available views are document dependent.

- *Zoom to Fit* : Fit the model to the Graphics window.

- *Zoom to Area* : Zoom to the areas you select with a bounding box.

- *Previous View* : Display the previous view.

- *Section View* : Display a cutaway of a part or assembly, using one or more cross section planes.

- *Dynamic Annotation Views* : Only available with SOLIDWORKS MBD. Control how annotations are displayed when you rotate a model.

The Orientation dialog has an option to display a view cube (in-context View Selector) with a live model preview. This helps the user to understand how each standard view orientates the model. With the view cube, you can access additional standard views. The views are easy to understand and they can be accessed simply by selecting a face on the cube.

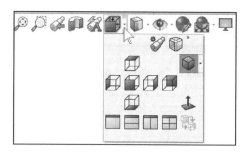

To activate the Orientation dialog box, press (Ctrl + spacebar) or click the View Orientation icon from the Heads-up View toolbar. The active model is displayed in the View Selector in an Isometric orientation (default view).

Click the View Selector icon in the Orientation dialog box to show or hide the in-context View Selector.

Press **Ctrl + spacebar** to activate the View Selector.

Press the **spacebar** to activate the Orientation dialog box.

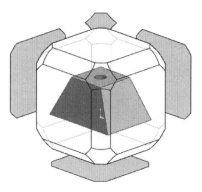

- *View Orientation box* : Select a view orientation or the number of viewports. The options are: *Top, Left, Front, Right, Back, Bottom, Single view, Two view - Horizontal, Two view - Vertical, Four view.* Click the drop-down arrow to access Axonometric views: *Isometric, Dimetric* and *Trimetric.*

- *Display Style* : Display the style for the active view. The options are: *Wireframe, Hidden Lines Visible, Hidden Lines Removed, Shaded, Shaded With Edges.*

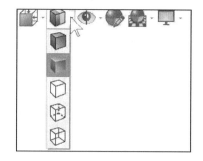

- *Hide/Show Items* 👁 ˅ : Select items to hide or show in the Graphics window. The available items are document dependent. Note the View Center of Mass ⊕ icon.

- *Edit Appearance* 🔵 : Edit the appearance of entities of the model.

- *Apply Scene* 🖼 ˅ : Apply a scene to an active part or assembly document. View the available options.

- *View Setting* 🖥 ˅ : Select the following settings: *RealView Graphics, Shadows In Shaded Mode, Ambient Occlusion, Perspective* and *Cartoon.*

- *Rotate view* ↻ : Rotate a drawing view. Input Drawing view angle and select the ability to update and rotate center marks with view.

- *3D Drawing View* 🔲 : Dynamically manipulate the drawing view in 3D to make a selection.

To display a grid for a part, click Options ⚙ ˅ , Document Properties tab. Click Grid/Snaps, check the Display grid box.

🔆 Add a custom view to the Heads-up View toolbar. Press the space key. The Orientation dialog box is displayed. Click the New View 🖌 tool. The Name View dialog box is displayed. Enter a new named view. Click OK.

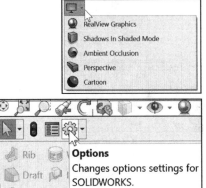

Use commands to display information about the triad or to change the position and orientation of the triad. Available commands depend on the triad's context.

🔆 Save space in the CommandManager, limit your CommandManager tabs. **Right-click** on a CommandManager tab. Click **Tabs**. View your options to display CommandManager tabs.

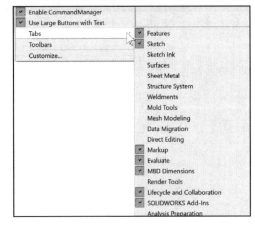

SOLIDWORKS CommandManager

The SOLIDWORKS CommandManager is a Context-sensitive toolbar. By default, it has toolbars embedded in it based on your active document type. When you click a tab below the CommandManager, it updates to display that toolbar. For example, if you click the Sketch tab, the Sketch toolbar is displayed.

The Lifecycle and Collaboration tab is displayed with all active tools if you purchased 3DEXPERIENCE education roles and are logged in.

Below is an illustrated CommandManager for a **Part** document. Tabs will vary depending on system setup and Add-ins.

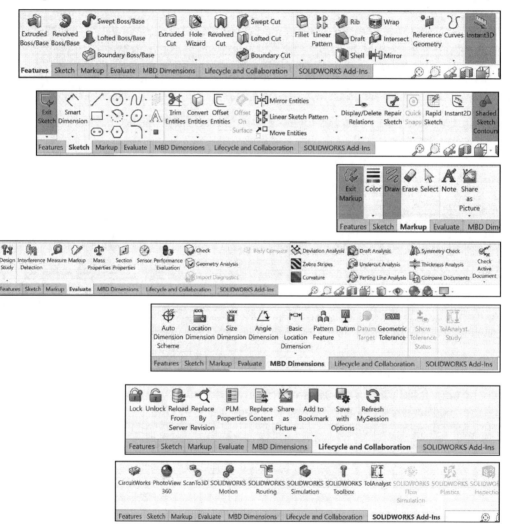

The SOLIDWORKS CommandManager is a Context-sensitive toolbar that automatically updates based on the toolbar you want to access. By default, it has toolbars embedded in it based on your active document type. The available tools are feature and document dependent.

Below is an illustrated CommandManager for a **Drawing** document. Tabs will vary depending on system setup and Add-ins.

To add a custom tab, right-click on a tab in the CommandManager and click Customize. You can also select to add a blank tab and populate it with custom tools from the Customize dialog box.

The SOLIDWORKS CommandManager is a Context-sensitive toolbar that automatically updates based on the toolbar you want to access. By default, it has toolbars embedded in it based on your active document type. The available tools are feature and document dependent.

Below is an illustrated CommandManager for an **Assembly** document. Tabs will vary depending on system setup and Add-ins.

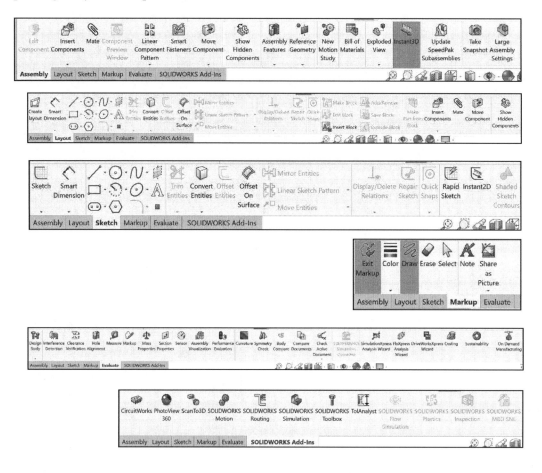

The Markup tab is displayed by default in the CommandManager on some systems. The Share as Picture option is available if you purchased 3DEXPERIENCE education roles and are logged in.

Float the CommandManager. Drag the Features, Sketch or any CommandManager tab. Drag the CommandManager anywhere on or outside the SOLIDWORKS window.

To dock the CommandManager, perform one of the following:

While dragging the CommandManager in the SOLIDWORKS window, move the pointer over a docking icon -

Dock above , Dock left , Dock right   and click the needed command.

Double-click the floating CommandManager to revert the CommandManager to the last docking position.

Collapse the CommandManager

Collapse the CommandManager to only display the available tabs. No tools are shown.

Click the Collapsed CommandManager arrow to the right of the active CommandManager tab as illustrated. Only the tabs are displayed.

To display the CommandManager tools back, click on a CommandManager tab. Click the Pin CommandManager as illustrated.

Part FeatureManager Design Tree

The Part FeatureManager consists of various tabs:

- *FeatureManager design tree* 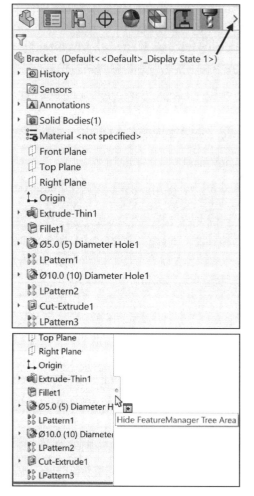 tab.

- *PropertyManager* tab.

- *ConfigurationManager* tab.

- *DimXpertManager* tab.

- *DisplayManager* tab.

- *CAM FeatureManager tree* tab.

- *CAM Operation tree* tab.

- *CAM Tools tree* tab.

Click the direction arrows to expand or collapse the FeatureManager design tree.

CommandManager and FeatureManager tabs and folder files will vary depending on your SOLIDWORKS applications and Add-ins.

Select the Hide/Show FeatureManager Area tab

as illustrated to enlarge the Graphics window for modeling.

The Sensors tool located in the FeatureManager monitors selected properties in a part or assembly and alerts you when values deviate from the specified limits. There are five sensor types: Simulation Data, Mass properties, Dimensions, Measurement and Costing Data.

Various commands provide the ability to control what is displayed in the FeatureManager design tree.

1. Show or Hide FeatureManager items.

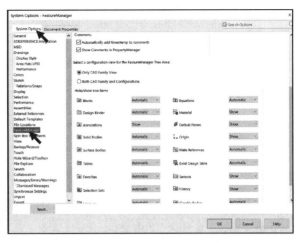

Click **Options** ⚙ ˙ from the Menu bar. Click **FeatureManager** from the System Options tab. Customize your FeatureManager from the Hide/Show tree Items dialog box.

2. Filter the FeatureManager design tree. Enter information in the filter field. You can filter by *Type of features, Feature names, Sketches, Folders, Mates, User-defined tags* and *Custom properties*.

Tags are keywords you can add to a SOLIDWORKS document to make them easier to filter and to search. The Tags ✎ icon is located in the bottom right corner of the Graphics window.

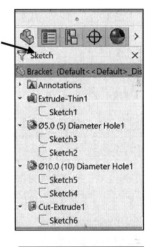

Collapse all items in the FeatureManager, **right-click** and select **Collapse items**, or press the **Shift + C** keys.

The FeatureManager design tree and the Graphics window are dynamically linked. Select sketches, features, drawing views, and construction geometry in either pane.

Split the FeatureManager design tree and either display two FeatureManager instances, or combine the FeatureManager design tree with the ConfigurationManager or PropertyManager.

Move between the FeatureManager design tree, PropertyManager, ConfigurationManager, DimXpertManager, DisplayManager and others by selecting the tab at the top of the menu.

Split

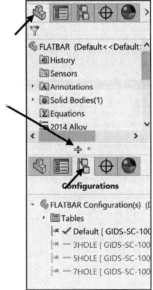

The ConfigurationManager ⊞ tab is located to the right of the PropertyManager ▤ tab. Use the ConfigurationManager to create, select and view multiple configurations of parts and assemblies.

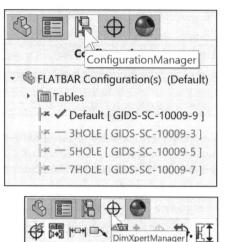

The icons in the ConfigurationManager denote whether the configuration was created manually or with a design table.

The DimXpertManager ⊕ tab provides the ability to insert dimensions and tolerances manually or automatically. The options are: **Auto Dimension Scheme** ⊕, **Auto Pair Tolerance** ⊞, **Basic, Location Dimension** |⊷|, **Basic Size Dimension** ⟋, **General Profile Tolerance** ⧉, **Show Tolerance Status** ±⊙, **Copy Scheme** ⊕, **Import Scheme** ⊕, **TolAnalyst Study** ⟀ and **Datum Target** ⊘.

🔆 TolAnalyst is available in SOLIDWORKS Premium.

Fly-out FeatureManager

The fly-out FeatureManager design tree provides the ability to view and select items in the PropertyManager and the FeatureManager design tree at the same time.

Throughout the book, you will select commands and command options from the drop-down menu, fly-out FeatureManager, Context toolbar, or from a SOLIDWORKS toolbar.

🔆 Another method for accessing a command is to use the accelerator key. Accelerator keys are special keystrokes, which activate the drop-down menu options. Some commands in the menu bar and items in the drop-down menus have an underlined character.

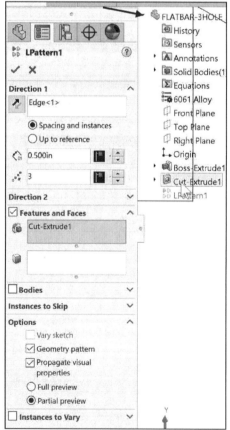

Task Pane

The Task Pane is displayed when a SOLIDWORKS session starts. You can show, hide, and reorder tabs in the Task Pane. You can also set a tab as the default so it appears when you open the Task Pane, pin or unpin to the default location. **Note**: The Task Pane icon order may be different.

The Task Pane contains the following default tabs:

- *SOLIDWORKS Resources* ⌂ .

- *Design Library* ▦ .

- *File Explorer* ▭ .

- *View Palette* ▦ .

- *Appearances, Scenes and Decals* ● .

- *Custom Properties* ▤ .

- *3DEXPERIENCE files on This PC* ▦ .

- *3DEXPERIENCE* ● .

💡 Additional tabs are displayed with Add-Ins.

SOLIDWORKS Resources

The SOLIDWORKS Resources ⌂ icon displays the following default selections:

- *Welcome to SOLIDWORKS.*

- *SOLIDWORKS Tools.*

- *Online Resources.*

- *Subscription Services.*

Other user interfaces are available during the initial software installation selection: *Machine Design, Mold Design, Consumer Products Design, etc.*

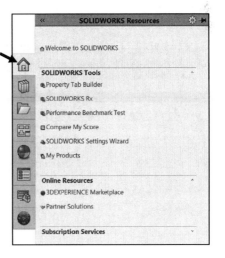

Design Library

The Design Library 📦 contains reusable parts, assemblies, and other elements including library features.

The Design Library tab contains default selections. Each default selection contains additional sub categories.

The default selections are:

- *Design Library.*

- *SOLIDWORKS Content (Internet access required).*

- *3D Components - PartSupply.*

- *Toolbox.*

🔆 Activate the SOLIDWORKS Toolbox. Click Tools, Add-Ins.., from the Main menu. Check the SOLIDWORKS Toolbox Library and SOLIDWORKS Toolbox Utilities box from the Add-ins dialog box or click SOLIDWORKS Toolbox from the SOLIDWORKS Add-Ins tab.

To access the Design Library folders in a non-network environment, click Add File Location 📦 and browse to the needed path. Paths may vary depending on your SOLIDWORKS version and window setup.

In a network environment, contact your IT department for system details.

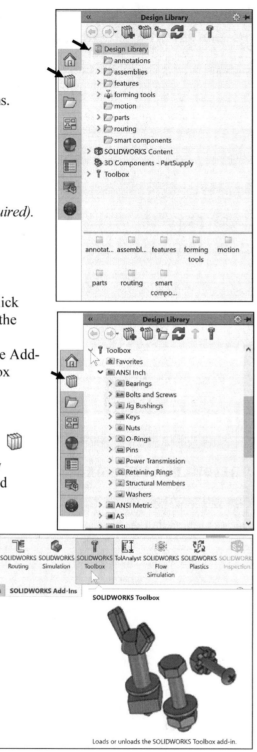

File Explorer

File Explorer tab duplicates Windows Explorer from your local computer and displays:

- *Open in SOLIDWORKS.*

- *Desktop.*

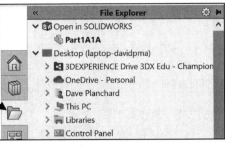

View Palette

The View Palette tab provides the ability to insert drawing views of an active document, or click the Browse button to locate the desired document.

Click and drag the view from the View Palette into an active drawing sheet to create a drawing view.

Appearances, Scenes, and Decals

Appearances, Scenes, and Decals tab provides a simplified way to display models in a photo-realistic setting using a library of Appearances, Scenes, and Decals.

An appearance defines the visual properties of a model, including color and texture. Appearances do not affect physical properties, which are defined by materials.

Scenes provide a visual backdrop behind a model. In SOLIDWORKS they provide reflections on the model. PhotoView 360 is an Add-in. Drag and drop a selected appearance, scene or decal on a feature, surface, part or assembly.

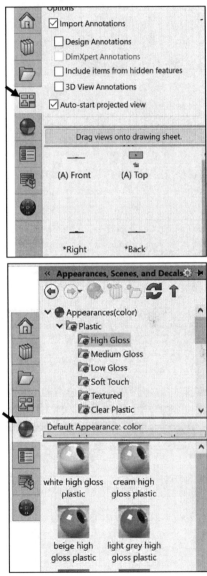

Custom Properties

The Custom Properties tab provides the ability to enter custom and configuration specific properties directly into SOLIDWORKS files.

3DEXPERIENCE files on This PC

The 3DEXPERIENCE files on This PC tab displays the models on your PC which were saved to the **3D**EXPERIENCE platform.

3DEXPERIENCE

The 3DEXPERIENCE tab provides access to the **3D**EXPERIENCE platform. The **3D**EXPERIENCE platform provides downloadable content (DLC) from a secure cloud location within your SOLIDWORKS session. The platform enables you to collaborate with students and teams on and off campus through a browser.

Think of the **3D**EXPERIENCE platform simply as a SOLIDWORKS Add-In on your desktop. Experience a single modeling environment. Access the platform within the SOLIDWORKS Task Pane.

Note: If you purchased 3DEXPERIENCE education roles, follow the below procedure.

Create and register your **3D**EXPERIENCE ID (credentials) to the platform. Go to https://solidworks.com/academic-cloud to create and register a new ID. **Note**: In an academic environment, it is strongly recommended to use your school's assigned email address.

Input the requested information.
Click **Register**.

You will receive the
3DEXPERIENCE platform
Invitation link via email. The
email you use must be the same
as your **3D**EXPERIENCE ID.

Click **Launch your
3DEXPERIENCE platform**
from the Invitation email.

Bookmark the address.

Store the Invitation email. It's
your unique link to the
3DEXPERIENCE platform.

Log in with your
3DEXPERIENCE
ID.

Dynamic
Reference
Visualization
(Parent/Child)

Dynamic Reference Visualization provides the ability
to view the parent/child relationships between items in
the FeatureManager design tree.

When you hover over a feature with references in the
FeatureManager design tree, arrows display showing
the relationships.

If a reference cannot be shown because a feature is not
expanded, the arrow points to the feature that contains
the reference and the actual reference appears in a text
box to the right of the arrow.

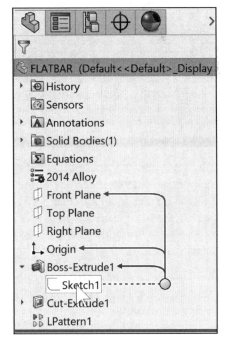

Use Dynamic reference visualization for a part, assembly and mates.

To display the Dynamic Reference Visualization, click **View**, **User Interface**, **Dynamic Reference Visualization Parent/Child)** from the Main menu bar.

Mouse Movements

A mouse typically has two buttons: a primary button (usually the left button) and a secondary button (usually the right button). Most mice also include a scroll wheel between the buttons to help you scroll through documents and to Zoom in, Zoom out and rotate models in SOLIDWORKS. It is highly recommended that you use a mouse with at least a Primary, Scroll and Secondary button.

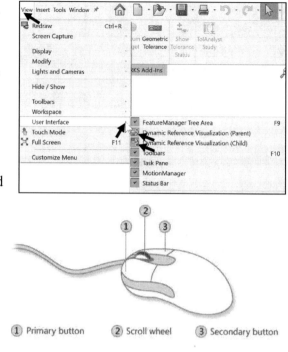

① Primary button   ② Scroll wheel   ③ Secondary button

Single-click

To click an item, point to the item on the screen, and then press and release the primary button (usually the left button). Clicking is most often used to select (mark) an item or open a menu. This is sometimes called single-clicking or left-clicking.

Double-click

To double-click an item, point to the item on the screen, and then click twice quickly. If the two clicks are spaced too far apart, they might be interpreted as two individual clicks rather than one double-click. Double-clicking is most often used to open items on your desktop. For example, you can start a program or open a folder by double-clicking its icon on the desktop.

Right-click

To right-click an item, point to the item on the screen, and then press and release the secondary button (usually the right button). Right-clicking an item usually displays a list of things you can do with the item. Right-click in the open Graphics window or on a command in SOLIDWORKS, and additional pop-up context is displayed.

Scroll wheel

Use the scroll wheel to zoom-in or to zoom-out of the Graphics window in SOLIDWORKS. To zoom-in, roll the wheel backward (toward you). To zoom-out, roll the wheel forward (away from you).

Saving SOLIDWORKS Documents as Previous version

Beginning with SOLIDWORKS 2024, you can save SOLIDWORKS parts, assemblies and drawings created or saved in the latest version of SOLIDWORKS as fully functional documents in a previous version of SOLIDWORKS.

You can save documents back to the previous two releases. Pack and Go also supports this functionality.

You can save SOLIDWORKS 2024 files as SOLIDWORKS 2023 or SOLIDWORKS 2022 versions.

This previous release compatibility lets you share files with others who use one of the two previous versions of SOLIDWORKS.

You cannot extend the previous release compatibility beyond those two releases.

Note: SOLIDWORKS users must have an active subscription license to access this functionality. 3DEXPERIENCE users are active subscribers by default.

The SOLIDWORKS (UI) is designed to make maximum use of the Graphics window for your model. Displayed toolbars and commands are kept to a minimum.

To open the Previous Release Check dialog box, click Tools, Evaluate, from the Main menu.

Summary

The SOLIDWORKS (UI) is designed to make maximum use of the Graphics window for your model. Displayed toolbars and commands are kept to a minimum.

The SOLIDWORKS User Interface and CommandManager consist of the following main options: Menu bar toolbar, Menu bar menu, Drop-down menus, Context toolbars, Consolidated fly-out menus, System feedback icons, Confirmation Corner and Heads-up View toolbar.

The Part CommandManager controls the display of tabs: *Features*, *Sketch*, Markup, *Evaluate*, *MBD Dimensions*, *Lifecycle and Collaboration* and various *SOLIDWORKS Add-Ins*.

The FeatureManager consists of various tabs:

- *FeatureManager design tree* tab.

- *PropertyManager* tab.

- *ConfigurationManager* tab.

- *DimXpertManager* tab.

- *DisplayManager* tab.

- *CAM FeatureManager tree* tab.

- *CAM Operation tree* tab.

- *CAM Tools tree* tab.

Click the direction arrows to expand or collapse the FeatureManager design tree.

CommandManager, **FeatureManager, and file folders will vary depending** on system set-up, Add-ins and 3DEXPERIENCE Roles.

You learned about creating a new SOLIDWORKS part and opening an existing SOLIDWORKS part along with using the Rollback bar to view the sketches and features.

If you modify a document property from an Overall drafting standard, a modify message is displayed as illustrated.

Overall drafting standard
ANSI-MODIFIED
Derived from: ANSI

Use the Search box, in the upper left corner of the Materials dialog box, to search through the entire materials library.

Templates are part, drawing and assembly documents which include user-defined parameters. Open a new part, drawing or assembly. Select a template for the new document.

In Chapter 2, establish a SOLIDWORKS session. Learn about 2D Sketching and 3D features.

Create a new part. Create the Wheel for the Fly Wheel sub-assembly.

Utilize the Fly Wheel sub-assembly in the final Stirling Engine assembly.

Apply the following sketch and feature tools: Circle, Line Centerline, Centerpoint Straight Slot, Mirror Entities, Extruded Boss, Extruded Cut, Revolved Boss, Circular Pattern, Hole Wizard and Fillet.

Incorporate design change into a part using proper design intent, along with applying multiple geometric relations: Coincident, Vertical, Horizontal, Tangent and Midpoint and feature and sketch modifications.

Utilize the Material, Mass Properties, Appearance and Measure tool on the Wheel.

Notes:

Chapter 2

2D Sketching, Features and Parts

Below are the desired outcomes and usage competencies based on the completion of Chapter 2.

Desired Outcomes:	Usage Competencies:

• Create a SOLIDWORKS Part document:
 	
  o Set drafting standard.
 	
  o Set units and precision.
 	
• Apply proper design intent.
 	
• Create a Wheel part.
 	
• Utilize the Wheel in the Fly Wheel sub-assembly.
 	
• Utilize the Fly Wheel in the final Stirling Engine assembly.
	• Aptitude to establish a SOLIDWORKS session
	
	• Set System and Document properties.
	
	• Knowledge of the following sketch and feature tools: Circle, Line, Centerline, Center point Straight Slot, Mirror Entities, Extruded Base, Revolved Boss, Extruded Cut, Circular Pattern, Hole Wizard and Fillet.
	
	• Ability to incorporate design change into the design.
	
	• Skill to insert multiple geometric relations.
	
	• Comprehend applying material and using the Mass Properties and Measure tools.

Notes:

Chapter 2 - 2D Sketching, Features and Parts

Chapter Objective

Establish a SOLIDWORKS session. Create a new part called Wheel with user defined System and Document properties. The Wheel part is used in the Fly Wheel sub-assembly. The sub-assembly is used in the final Stirling Engine assembly.

On the completion of this chapter, you will be able to:

- Establish a SOLIDWORKS session.

- Set user defined System and Document properties.

- Recognize the default Sketch Planes in the Part FeatureManager.

- Select the correct Sketch plane and orientation for the Wheel.

- Utilize the following Sketch tools: Circle, Line, Centerline, Center point Straight Slot and Mirror Entities.

- Comprehend and apply proper design intent.

- Edit a Sketch and Sketch Plane.

- Insert the following Geometric relations: Coincident, Vertical, Horizontal, Tangent and Midpoint.

- Create and modify the following features: Extruded Base, Revolved Boss, Extruded Cut, Circular Pattern, Hole Wizard and Fillet.

- Apply Material and utilize the Mass Properties and Measure tool to verify a design change.

- Utilize and apply the Appearance tool.

Activity: Start a SOLIDWORKS Session. Create a New Part Document.

Start a SOLIDWORKS session.

1) Double-click the **SOLIDWORKS 2024 icon** from the Desktop.

2) **Close** the Welcome - SOLIDWORKS dialog box. The SOLIDWORKS program is displayed.

Create a new part document.

3) Click **New** ⬜ from the Menu bar or click **File**, **New** from the Menu bar menu. The New SOLIDWORKS Document dialog box is displayed. Advanced mode is used in this book.

4) If needed, click the **Advanced** tab. The below New SOLIDWORKS Document box is displayed.

5) Click the **Templates** tab.

6) Double-click **Part** from the New SOLIDWORKS Document dialog box. Part 1 is displayed in the FeatureManager.

The first system default Part filename is Part1. The system attaches the .sldprt suffix to the created part. The second created part in the same session increments to the filename Part2.

Display the origin.

7) Click **View**, **Hide/Show**, **Origins** from the Main menu. The origin is displayed in the Graphics window.

The Part Origin ⊥ is displayed in blue in the center of the Graphics window.

The Origin represents the intersection of the three default reference planes: *Front Plane*, *Top Plane* and *Right Plane*.

The positive X-axis is horizontal and points to the right of the Origin in the Front view.

The positive Y-axis is vertical and points upward in the Front view. The FeatureManager contains a list of features, reference geometry, and settings utilized in the part.

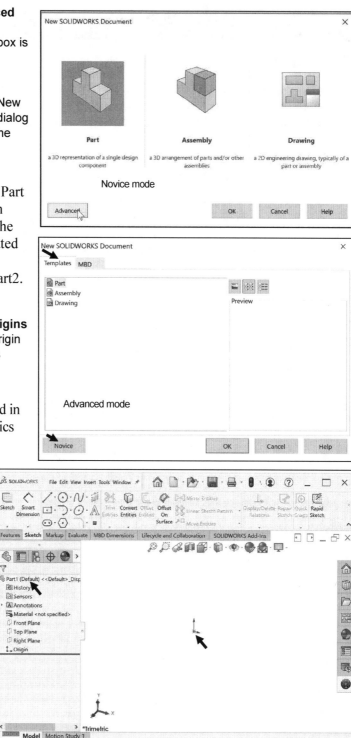

Activity: Set Document Properties.

Set Document Properties. Set overall drafting standard.

8) Click **Options** ⚙️ from the Menu bar toolbar. The System Options General dialog box is displayed.

9) Click the **Document Properties** tab.

10) Select **ANSI** from the Overall drafting standard drop-down menu.

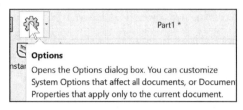

🔆 Various detailing options are available depending on the selected standard.

The Overall drafting standard determines the display of dimension text, arrows, symbols, and spacing. Units are the measurement of physical quantities. Millimeter dimensioning and decimal inch dimensioning are the two most common unit types specified for engineering parts and drawings.

Set Document Properties. Set units and precision.

11) Click the **Units** folder.

12) Click **MMGS** (millimeter, gram, second) for Unit system.

13) Select **.12** (two decimal places) for Length basic units.

14) Click **OK** from the Document Properties - Units dialog box. The Part FeatureManager is displayed.

🔆 The origin ↳ represents the intersection of the Front, Top and Right planes.

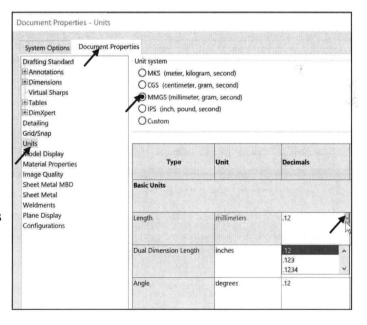

2D Sketching - Identify the Correct Sketch Plane

Most SOLIDWORKS features start with a 2D sketch. Sketches are the foundation for creating features. SOLIDWORKS provides the ability to create either 2D or 3D sketches.

A 2D sketch is limited to a flat 2D sketch plane located on a reference plane, face or a created plane. 3D sketches are very useful in creating sketch geometry that does not lie on an existing or easily defined plane.

Does it matter what plane you start the base 2D sketch on? Yes. When you create a new part or assembly, the three default planes are aligned with specific views. The plane you select for your first sketch determines the orientation of the part. Selecting the correct plane to start your model is very important.

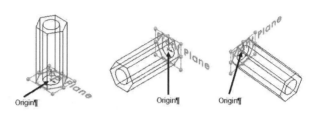

Sketch States

Sketches can exist in any of five states. The state of the sketch is displayed in the status bar at the bottom of the SOLIDWORKS window. The following are the five sketch states in SOLIDWORKS:

1. *Under Defined.* Inadequate definition of the sketch (blue). The FeatureManager displays a minus (-) symbol before the sketch name.

2. *Fully Defined.* Complete information (black). The FeatureManager displays no symbol before the sketch name.

3. *Over Defined.* Duplicate dimensions and or relations (orange - red). The FeatureManager displays a (+) symbol before the sketch name. The What's Wrong dialog box is displayed.

4. *Invalid Solution Found.* Your sketch is solved but results in invalid geometry, for example, a zero length line, zero radius arc or a self-intersecting spline (yellow).

5. *No Solution Found.* Indicates sketch geometry that cannot be resolved (Brown).

Color indicates the state of the individual Sketch entities.

In SOLIDWORKS, it is not necessary to fully dimension or define sketches before you use them to create features. You should fully define sketches before you consider the part finished for manufacturing.

Activity: Create the Base Sketch for the First Feature of the Wheel.

In SOLIDWORKS, a 2D profile is called a sketch. The Base sketch is the first sketch in the feature. A sketch requires a sketch plane and a 2D profile. The sketch in this example uses the Front Plane. The 2D profile is a circle.

A Geometric relationship and a dimension define the exact size and location of the center point of the circle relative to the origin.

The center of the circle is Coincident with the origin. The origin is displayed in red.

Create the Base Sketch.

15) Right-click **Front Plane** from the FeatureManager. The Context toolbar is displayed. This is your Sketch plane. The Sketch toolbar is displayed. Front Plane is your Sketch plane.

16) Click **Sketch** ⌐ from the Context toolbar as illustrated.

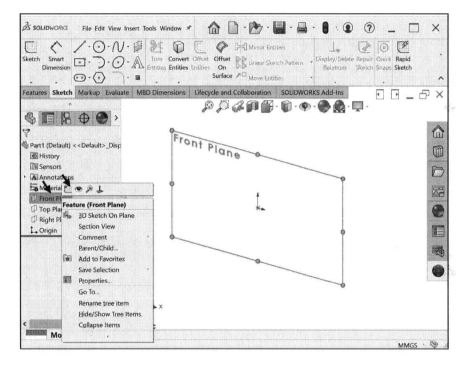

💡 CommandManager and FeatureManager tabs and folder files will vary depending on system setup and Add-ins.

💡 If needed, click View, Hide/Show from the Main menu bar. Click Sketch Relations to display relations in the Graphics window.

The Front Plane rotates normal to the Sketch Plane.

17)   Click the **Circle** ⊙ tool from the Sketch toolbar. The Circle PropertyManager is displayed. The sketch opens in the Front view. The mouse pointer displays the Circle symbol icon. The Front Plane feedback indicates the current Sketch plane.

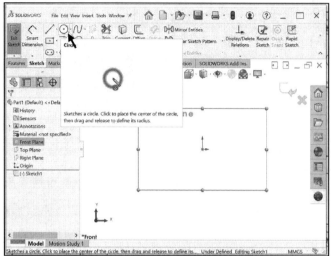

The Circle-based tool uses a Consolidated Circle PropertyManager. The SOLIDWORKS application defaults to the last used tool type.

18)   Move the **mouse pointer** into the Graphics window. The cursor displays the Circle symbol icon.

19)   Click the **Origin** ↳ in the Graphics window. This is the first point of the circle. It is very important that you always reference the sketch to the origin. This helps to fully define the sketch. The cursor displays a Coincident relation ▓ to the origin.

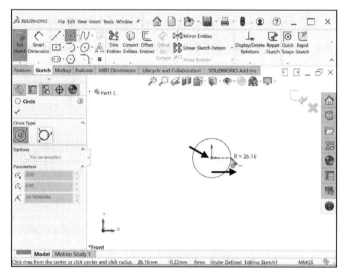

20)   Drag the **mouse pointer** to the right of the origin (approximately 25mm) to create the circle as illustrated. The center point of the circle is positioned at the origin.

21)   Click a **position** to create the circle. The sketch is under defined and is displayed in blue.

If needed, view the sketch relations in the Graphics window.

22) Click **View**, **Hide/Show**, **Sketch Relations** from the Main Menu bar. The sketch relation (Coincident ⊠) is displayed in the Graphics window.

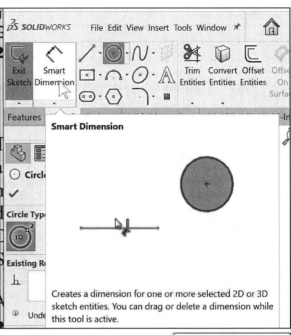

The diameter of the circle is displayed above the mouse pointer as you drag the mouse pointer up and to the right. The diameter displays different values. Define the exact dimension with the Smart dimension tool.

Add a dimension to fully define the sketch.

23) Click the **Smart Dimension** ✎ tool from the Sketch toolbar. The cursor displays the Smart Dimension ✎ icon.

24) Click the **circumference** of the circle.

25) Click a **position** diagonally above and to the right of the circle in the Graphics window.

26) Enter **20**mm in the Modify dialog box.

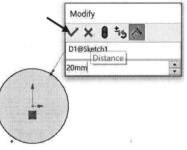

27) Click the **Green Check mark** ✔ in the Modify dialog box. The diameter of the circle is 20mm. The sketch is fully defined and is displayed in black.

💡 Add relations then dimensions. This keeps the user from having too many unnecessary dimensions. This also helps to show the design intent of the model.

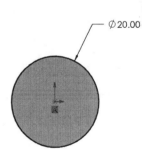

Features

- Features are geometry building blocks.

- Features add or remove material.

- Features are created from sketched profiles or from edges and faces of existing geometry.

What is a Base feature? The Base feature for the Wheel (Extruded Base) is the first feature that is created. The Base feature is the foundation of the part. Keep the Base feature simple.

Activity: Create the First Feature of the Wheel - Extruded Base.

Create the first feature of the Wheel. Extrude the sketch.

28) Click the **Features** tab from the CommandManager. The Features toolbar is displayed.

29) Click **Extruded Boss/Base** from the Features toolbar. The Boss-Extrude PropertyManager is displayed. Blind is the default End Condition in Direction 1. The extruded sketch is previewed in a Trimetric view. The preview displays the direction of the Extrude feature.

30) Select **Mid Plane** for End Condition in Direction 1. Note your End Condition options.

31) Enter **20**mm for Depth in Direction 1. Accept the default conditions.

32) Click **OK** ✔ from the Boss-Extrude PropertyManager. Boss-Extrude1 is displayed in the FeatureManager.

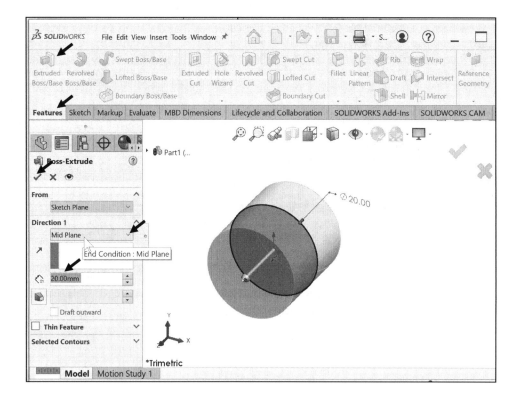

Fit the model to the Graphics window.

33)   Press the **f** key.

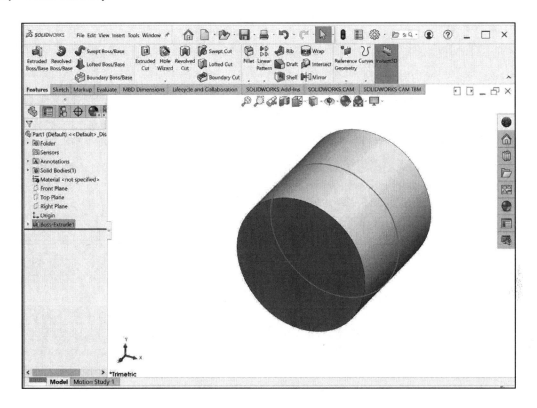

Design Intent

What is design intent? All designs are created for a purpose. Design intent is the intellectual arrangements of features and dimensions of a design. Design intent governs the relationship between sketches in a feature, features in a part and parts in an assembly.

The SOLIDWORKS definition of design intent is the process in which the model is developed to accept future modifications. Models behave differently when design changes occur.

Design for change. Utilize geometry for symmetry, reuse common features, and reuse common parts. Build change into the following areas that you create: sketch, feature, part, assembly and drawing.

Start the translation of the initial design functional and geometric requirements into SOLIDWORKS features.

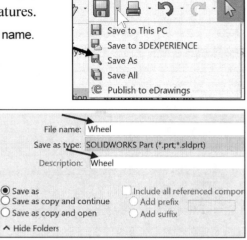

Save the part in the SOLIDWORKS 2024 folder. Enter name. Enter description.

34)   Click **Save As** from the Drop-down Menu bar.

35)   Double-click the **SOLIDWORKS 2024**. folder. The folder was downloaded in Chapter 1.

36)   Enter **Wheel** for File name.

37)   Enter **Wheel** for Description.

38)   Click **Save** from the Save As dialog box. The Wheel FeatureManager is displayed.

39)   **Expand** Boss-Extrude1 from the FeatureManager as illustrated. Sketch1 is fully defined.

A fully defined sketch has complete information (manufacturing and inspection) and is displayed in black.

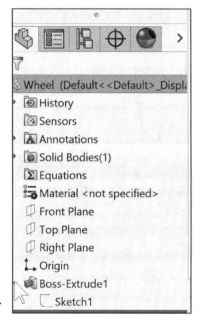

Display an Isometric view of the model. Press the **space bar** to display the Orientation dialog box. Click the **Isometric view** icon.

Organize parts into folders. The folder for this chapter is named SOLIDWORKS 2024. All documents for this book are saved in the SOLIDWORKS 2024 folder.

FeatureManager tabs and folder files will vary depending on system setup and SOLIDWORKS Add-ins.

In the next section, edit the Sketch and Sketch plane.

Activity: Edit the Base Sketch and Sketch Plane.

Edit the Base Sketch from 20mm to 40mm.

40) Right-click **Sketch1** in the FeatureMananger.

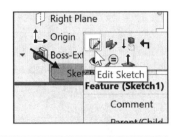

41) Click **Edit Sketch** from the Context toolbar.

42) Double-click **20** in the Graphics window. The Modify dialog box is displayed.

43) Enter **40**mm in the Modify dialog box.

44) Click the **Green Check mark** in the Modify dialog box. The diameter of the circle is 40mm. View the results.

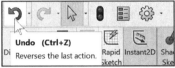

Undo the modification from 20mm to 40mm.

45) Click **Undo** from the Main menu as illustrated.

46) Click **Exit Sketch** from the Sketch toolbar to return to the original model.

Edit the Base Sketch Plane.

47) Right-click **Sketch1** in the FeatureMananger.

48) Click **Edit Sketch Plane** from the Context toolbar. The Sketch Plane PropertyManager is displayed.

49) **Expand** the fly-out Wheel FeatureManager as illustrated.

50) Click **Top Plane** from the fly-out Wheel FeatureManager. Top Plane is displayed in the Sketch Plane/Face box.

51) Click **OK** from the Sketch Plane PropertyManager.

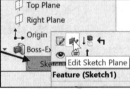

52) **View** the model orientation in the Graphics window.

Return to the original orientation.

53) Click **Undo**. The original orientation is displayed (Front Plane).

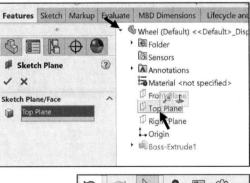

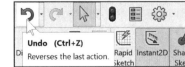

Display Modes, View Modes, View Tools, and Appearances

Access the display modes, view modes, view tools and appearances from the Standard Views toolbar and the Heads-up View toolbar. Apply these tools to display the required modes in your document.

The Apply scene 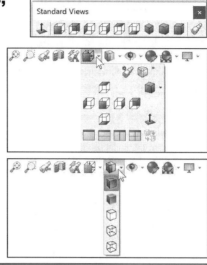 tool adds the likeness of a material to a model in the Graphics window without adding the physical properties of the material. Select a scene from the drop-down menu.

Activity: Create the Second Feature of the Wheel - Revolved Boss.

Create the second feature on the Right Plane.

54)　Right-click **Right Plane** from the FeatureMananger.

55)　Click **Sketch** from the Context toolbar.

Display a Right view and Zoom out.

56)　Click **Right view** from the Heads-up View toolbar.

57)　Press the **z** key approximately five times to zoom out.

Display the origin.

58)　If needed, click **View**, **Hide/Show**, **Origins** from the Main menu. The origin is displayed in the Graphics window.

Create the profile of the Wheel.

59)　Click the **Centerline** Sketch tool from the drop-down menu. The Insert Line PropertyManager is displayed.

60)　Click the **midpoint** (it is very important that you select the midpoint) of the top edge of Boss-Extrude1 as illustrated. A Midpoint relation is displayed.

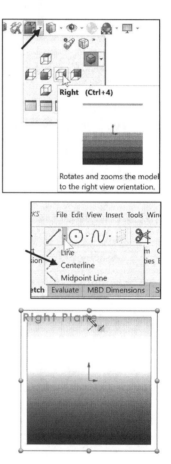

61) Sketch a **vertical centerline** from the midpoint as illustrated. Approximately 65mm.

Deselect the Centerline Sketch tool.

62) Right-click a **position** in the Graphics window. The Context toolbar is displayed.

63) Click **Select**. The mouse pointer displays the Select icon and the Centerline Sketch tool is deactivated. The sketch is displayed in black.

Select the Line Sketch tool.

64) Click the **Line** Sketch tool. The Insert Line PropertyManager is displayed.

65) Click the **midpoint** of the top edge of Boss-Extrude1.

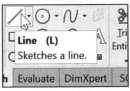

66) Click a **position** to the right as illustrated. Approximately 5mm long. A horizontal relation is displayed.

67) Click a **position** directly above as illustrated. Approximately 30mm long. A vertical relation is displayed.

68) Sketch an **angle** (*30degrees or LESS*) **line** approximately 5mm long as illustrated. This is a very important step.

69) Sketch a **vertical line** as illustrated. Approximately 5mm long.

70) Sketch a **horizontal line** to complete the sketch.

71) Click the **Centerline** to end the sketch. Note the automatic geometric relations in the sketch.

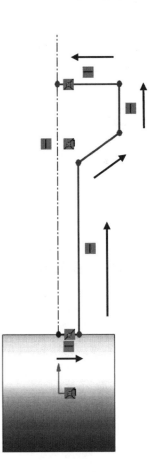

Deselect the Line Sketch tool.

72) Right-click a **position** in the Graphics window. The Context toolbar is displayed.

73) Click **Select**. The mouse pointer displays the Select icon and the Centerline Sketch tool is deactivated.

In SOLIDWORKS, relations between sketch entities and model geometry, in either 2D or 3D sketches, are an important means of building in design intent. As you sketch, allow the SOLIDWORKS application to automatically add relations.

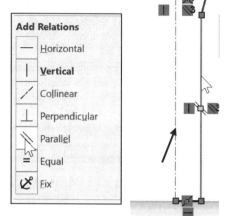

Automatic relations rely on Inferencing, Pointer display, Sketch Snaps and Quick Snaps. After you sketch, manually add relations if needed and the dimensions to fully define the sketch.

To manually add a geometric relation, click a sketch entity. Hold the Ctrl key down. Click the second sketch entity. The Properties PropertyManager is displayed. Click the needed geometric relation. In this case, a Parallel relation was created between the centerline and the first vertical line.

Mirror the open sketch profile.

74) Click the **Mirror Entities** ⊬ Sketch tool. The Mirror PropertyManager is displayed.

75) Click inside the **Mirror about** box.

76) Click the **Centerline** in the Graphics window. Line1 is displayed in the Mirror about box.

77) Click inside the **Entities to mirror** box.

Window-select the sketch entities.
78) Click in the **upper left corner** of the Graphics Window as illustrated.

79) **Drag** the mouse pointer to the lower right corner as illustrated.

80) **Release** the mouse button. The 5 sketch entities are displayed in the Entities to mirror box.

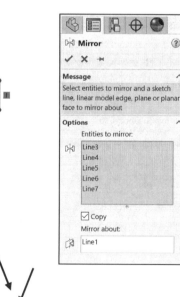

81) Click **OK** ✅ from the Mirror PropertyManager.

Deselect the Sketch Relations.

82) Click **View**, **Hide/Show**, uncheck **Sketch Relations** from the Main menu. View the results in the Graphics window.

Dimension the Sketch.

83) Click **Smart Dimension** ✎ from the Sketch toolbar.

The pointer displays the dimension symbol ✎ icon.

Dimension the angle.

84) Click the **two** illustrated lines. A dimension value is displayed.

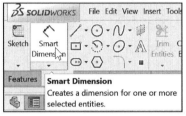

85) Click a **position** to the left as illustrated. The Modify dialog box is displayed. The Smart Dimension tool uses the Smart Dimension PropertyManager. The PropertyManager provides the ability to select three tabs. Each tab provides a separate menu.

86) Enter **30**deg in the Modify dialog box. The Dimension Modify dialog box provides the ability to select a unit drop-down menu to modify units in a sketch or feature from the document properties.

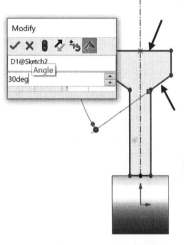

87) Click the **Green Check mark** ✅ in the Modify dialog box.

88) **Zoom in** on the bottom horizontal line.

Dimension the bottom horizontal line.

89) Click the **bottom horizontal line** as illustrated.

90) Click a position **below** the model.

91) Enter **4**mm in the Modify dialog box.

92) Click the **Green Check mark** ✅ in the Modify dialog box.

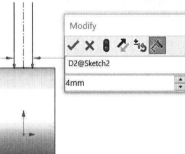

Fit the model to the Graphics window.

93) Press the **f** key.

Dimension the overall width of the top horizontal line.
94) Click the **top horizontal line** as illustrated.

95) Click a position **above** the model.

96) Enter **10**mm in the Modify dialog box.

97) Click the **Green Check mark** in the Modify dialog box.

Dimension the first vertical line to the right.
98) Click the **first vertical line**.

99) Click a position to the **right** of the line.

100) Enter **30**mm in the Modify dialog box.

101) Click the **Green check mark** in the Modify dialog box.

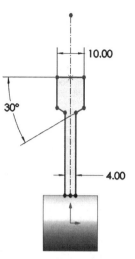

Dimension the overall height.
102) Click the **top horizontal line**.

103) Click the **bottom horizontal line**.

104) Click a position to the **right** of the line as illustrated.

105) Enter **35**mm in the Modify dialog box.

106) Click the **Green check mark** in the Modify dialog box. The sketch is fully defined and is displayed in black.

Display an Isometric view.
107) Click **Isometric view** from the Heads-up View toolbar.

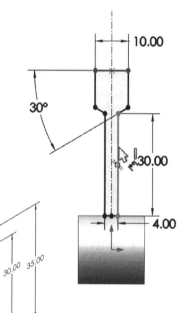

Create a Revolved Boss Feature.
108) Click the **Features** tab.

109) Click **Revolved Boss/Base** 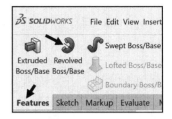 from the Features toolbar. The Revolve PropertyManager is displayed.

110) Right-click inside the **Axis of Revolution** box.

111) Click **Delete**.

Display the Temporary Axes.
112) Click **View Temporary Axes** from the Hide/Show Items in the Heads-up toolbar.

113) Click the **Temporary Axis** in the Graphics window as illustrated. Axis 1 is displayed in the Axis of Revolution box.

114) Enter **360** for Direction 1 Angle.

115) Click **OK** from the Revolve PropertyManager. View the results in the FeatureManager and in the Graphics window. Revolve1 is created.

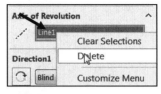

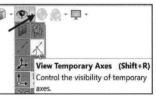

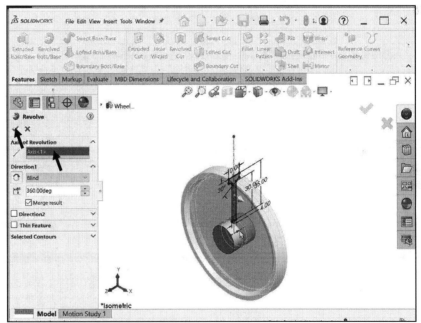

Deactivate the Temporary Axes.
116) Uncheck **View Temporary Axes** from the Hide/Show Items in the Heads-up toolbar.

117) Click **inside** the Graphics window.

118) Expand Revolve1 in the FeatureManager. Sketch2 is fully defined.

Activity: Create the Third Feature of the Wheel - Extruded Cut (Slot).

Display a Front view.

119) Click **Front view** from the Heads-up View toolbar.

Utilize the Centerpoint Straight Slot Sketch tool on the front face of the Wheel.

120) Right-click the **front face** of the Wheel as illustrated. The Context toolbar is displayed.

121) Click **Sketch** from the Context toolbar.

122) Click **Centerpoint Straight Slot** from the Consolidated Sketch toolbar. The Slot PropertyManager is displayed.

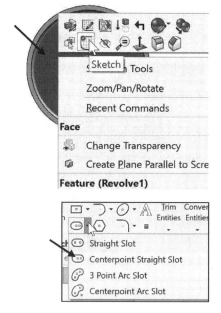

123) Click a **position** directly above the origin (first point).

124) Click a **position** directly above the first point (second point) as illustrated.

125) Click a **position** directly to the right of the second point (third point) as illustrated. The Centerpoint Straight Slot requires three points. Note: Add any needed geometric relations.

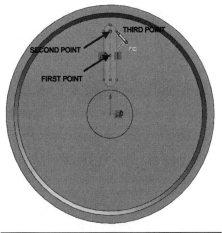

Utilize Construction geometry. Create a construction circle to locate the position of the center of the Slot relative to the Origin of the Wheel.

Locate the center of the slot. Create a Construction Circle.

126) Click the **Circle** ⊙ Sketch tool.

127) Click the **Origin** of the Wheel.

128) Click the **First point** of the slot.

129) Check the **For construction** box as illustrated.

130) Click **OK** ✔ from the Circle PropertyManager. A Construction circle is displayed.

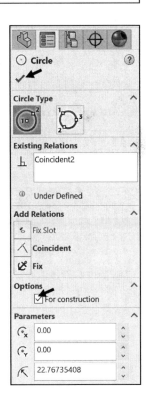

Insert dimensions to fully define the sketch.

131) Click **Smart Dimension** ✎ from the Sketch toolbar. The pointer displays the dimension symbol ✎ icon.

132) Enter the **dimensions** as illustrated. The sketch is fully defined and is displayed in black.

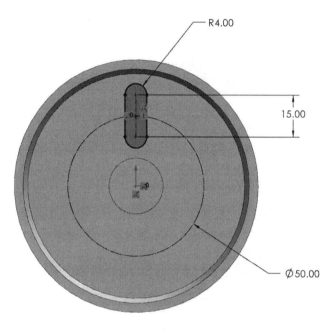

Create the Extruded Cut feature. The Extruded Cut feature is the Seed feature for the Circular Pattern. A Seed feature is the original feature for any type of pattern (linear, circular, sketch driven, curve driven, fill or mirrored pattern).

Create the Extruded Cut Feature.
133) Click the **Features** tab.

134) Click **Extruded Cut** 📄 from the Features toolbar. The Cut-Extrude PropertyManager is displayed.

135) Select **Through All** for End Condition. View the End Condition options.

136) Click **OK** ✔ from the Cut-Extrude PropertyManager. View the results in the Graphics window. Cut-Extrude1 is highlighted in the FeatureManager.

💡 Use various end conditions to ensure proper design intent.

View Sketch3.

137) Expand Cut-Extrude1 in the Part FeatureMananger. Sketch3 is fully defined.

Cut-Extrude1 is the seed feature for the Circular Pattern of slots on the Wheel. In the next section, create the Circular Pattern feature.

Save the model.

138) Click **Save** 🖫.

🔅 CommandManager and FeatureManager tabs and folder files will vary depending on system setup and Add-ins.

Select the Seed Feature for the Circular Pattern.

139) Click **Cut-Extrude1** in the FeatureManager. Cut-Extrude1 is highlighted.

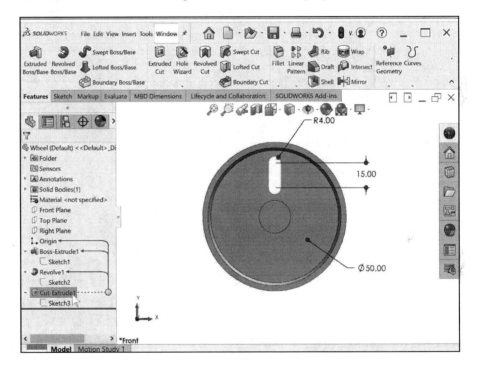

Activity: Create the Fourth Feature of the Wheel - Circular Pattern.

Display an Isometric view.

140) Click **Isometric view** from the Heads-up View toolbar.

Create the Circular Pattern feature.

141) Click **Circular Pattern** from the Features toolbar. The Circular Pattern PropertyManager is displayed. Cut-Extrude1 is selected and displayed in the Features to Pattern box.

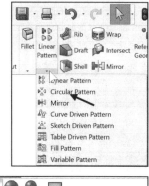

Display the Temporary Axes.

142) Click **View Temporary Axes** from the Hide/Show Items in the Heads-up toolbar.

143) Click the **Center Temporary Axis** in the Graphics window. Axis <1> is displayed in the Pattern Axis box.

144) Check the **Equal spacing** box.

145) Enter **8** for Number of Instances.

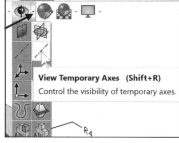

146) Click **OK** from the Circular Pattern PropertyManager. CirPattern1 is displayed in the FeatureManager. View the circular pattern of slots in the model.

Deactivate the Temporary Axes.

147) Uncheck **View Temporary Axes** from the Hide/Show Items in the Heads-up toolbar.

Save the model.

148) Click **Save**.

The Wheel requires a hole. Apply the Hole Wizard feature.

The Hole Wizard creates simple and complex Hole features by stepping through a series of options to define the hole type and hole position. The Hole Wizard requires a face or Sketch plane to position the Hole feature. Select the Front face of the hub.

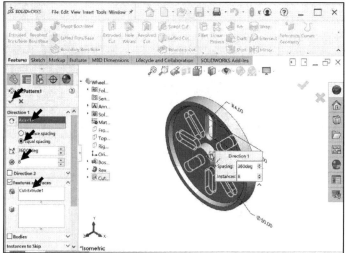

Activity: Create the Fifth Feature of the Wheel - Hole.

Display a Front view.

149) Click **Front view** 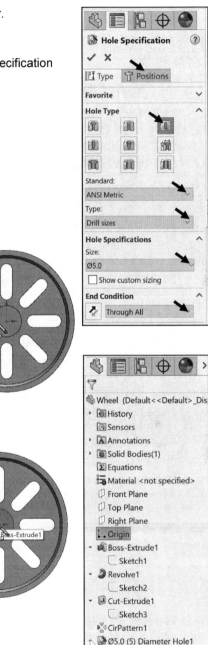 from the Heads-up View toolbar.

Insert the hole using the Hole Wizard tool.

150) Click the **Hole Wizard** Feature tool. The Hole Specification PropertyManager is displayed.

151) Click the **Type** tab.

152) Click the **Hole** icon for Hole Type.

153) Select **ANSI Metric** for Standard.

154) Select **Drill sizes** for Type.

155) Select Ø**5.0** for Size.

156) Select **Through All** for End Condition. Accept the default settings.

157) Click the **Positions** tab.

Select the Sketch plane for the Hole Wizard.

158) Click the **front face** of Base-Extrude1 as illustrated. Do not click the origin.

The Point tool is selected.

159) Click the **Origin** to place the center point of the hole. Right-click **Select** in the Graphics window to de-select the Point tool.

160) Click **OK** from the Hole Position PropertyManager. View the results.

Fit the WHEEL to the Graphics window.

161) Press the **f** key.

The Hole Wizard tool creates two sketches. The first sketch is the hole profile. The second sketch is to fully define the hole location relative to the origin.

Fillet/Round creates a rounded internal or external face on the part. You can fillet all edges of a face, selected sets of faces, selected edges, or edge loops.

In the next section, apply the Fillet feature to the Wheel. Create a Constant Size Fillet feature to 6 edges and 2 faces with a 2mm radius.

Activity: Create the Sixth Feature of the Wheel - Constant Size Fillet Feature.

Create a Constant Size Fillet Feature on six edges and two faces.

162) Click the **Fillet** 🖾 Feature from the Features toolbar. The Fillet PropertyManager is displayed.

163) Click the **Manual** tab.

164) Click **Constant Size Fillet** for Fillet Type.

165) Enter **2**mm for Radius in the Fillet Parameters box.

166) Click the **front narrow face** of Revolve 1 as illustrated. Face1 is displayed in the Items To Fillet box.

167) Click the **top front edge** of Boss-Extrude1 in the Graphics window. Edge1 is displayed in the Items To Fillet box.

168) Click the **bottom front edge** of the hub (Revolve1). Edge2 is displayed in the Items To Fillet box. Note the Fillet pop-up menu.

169) Click the **inside circuluar edge** of Revolve1. Edge3 is displayed in the Items To Fillet box.

170) Rotate the **Wheel** and select the **three edges** and a **face** as above. Edge4, Edge5, Edge6 and Face2 are dispayed in the Items To Fillet box. Zoom-in on the selected edges.

171) Click **OK** ✔ from the Fillet FeatureManager. Fillet1 is displayed in the FeatureManager.

172) **Rotate** the Wheel and view the results in the Graphics window.

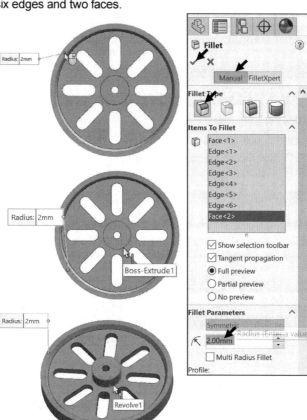

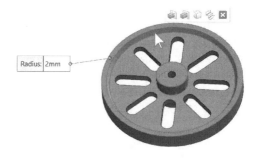

Save the model.

173) Click **Save** 💾.

Display a Trimetric view.

174) Click **Trimetric view** 🧊 from the Heads-up View toolbar.

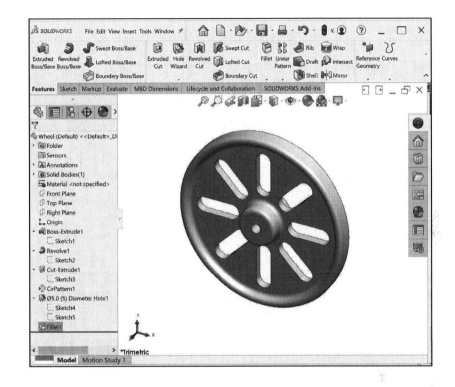

The Material dialog box helps you manage physical materials. You can work with pre-defined materials, create custom materials, apply materials to parts, and manage favorites.

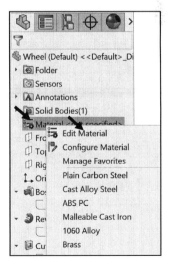

The left side of the Material dialog box contains a tree of available material types and materials. Tabs on the right display information about the selected material. If SOLIDWORKS Simulation is added in, additional tabs are displayed.

In the next section, apply material (6061 Alloy) to the Wheel.

Activity: Apply Material to the Wheel.

Apply Material.

175) Right-click **Material** from the FeatureManager.

176) Click **Edit Material**. The Material dialog box is displayed. View your options and material choices.

177) **Expand** the Aluminum Alloy folder.

178) Click **6061 Alloy**.

179) Click **Apply**.

180) Click **Close**. The Material is applied to the model.

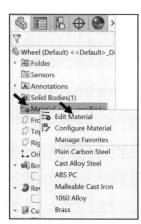

 Use the Search feature in the Material dialog box, to quickly located material type.

Activity: View the Mass Properties of the Wheel.

View the calculated mass properties of the model. You can assign values for mass, center of mass, and moments of inertia to override the calculated values.

View the Mass Properties of the Wheel.
181) Click the **Evaluate** tab from the CommandManager.

182) Click the **Mass Properties** icon. The Mass Properties dialog box is displayed. View the results. The total mass is 83.27 grams. The numbers represent the document properties (2 decimal places). Your number may differ by 1- 2% due to fillets.

Modify the precision in the Mass Properties dialog box to view Density.
183) Click the **Options** button.

184) Check the **Use custom settings** button.

185) Select **4** for Decimal places.

186) Click **OK** from the Mass/Section Property Options dialog box. Density = .0027 grams per cubic millimeters.

Close the Mass Properties dialog box.
187) Click **Close** ☒.

Display an Isometric view - Shaded With Edges.
188) Press the **space bar** to display the Orientation dialog box.

189) Click the **Isometric view** icon. You can also access the Isometric view tool from the Heads-up View toolbar.

190) Click the **Shaded With Edges** icon.

Save the Wheel.
191) Click **Save** 💾.

In the next section, modify the Circular Pattern feature. Decrease the number of instances from 8 to 4. Compare the Mass Properties of the modified model.

Mass properties of Wheel
  Configuration: Default
  Coordinate system: -- default --

Density = 0.00 grams per cubic millimeter

Mass = 83.27 grams

Volume = 30842.24 cubic millimeters

Surface area = 15835.01 square millimeters

Center of mass: (millimeters)
  X = 0.00
  Y = 0.00
  Z = 0.00

Mass/Section Property Options   ✕

Units
☐ Scientific Notation
○ Use document settings
● Use custom settings
    Length: Decimal places:
    Millimeters   4
    Mass:
    grams
    Per unit volume:
    millimeters^3

Activity: Modify the Circular Pattern. Decrease the Number of Instances.

Modify the CirPattern1 feature.

192) Right-click **CirPattern1** from the FeatureManager.

193) Click **Edit Feature** from the Content toolbar. The CirPattern1 PropertyManager is displayed.

194) Enter **4** for Number of Instances.

195) Click **OK** ✅ from the CirPattern PropertyManager. View the results in the PropertyManager.

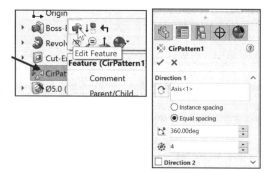

Activity: View the New Mass Properties of the Wheel

196) Click the **Evaluate** tab from the CommandManager.

197) Click the **Mass Properties** icon. The Mass Properties dialog box is displayed. View the results. The total mass is 90.63 grams vs. 83.27 grams. You removed 4 slots (instances) from the model.

Close the Mass Properties dialog box.

198) Click **Close** ❌.

Activity: Modify the Circular Pattern. Return to the Original Instances.

Modify the CirPattern1 feature.

199) Right-click **CirPattern1** from the FeatureManager.

200) Click **Edit Feature** from the Content toolbar. The CirPattern1 PropertyManager is displayed.

201) Enter **8** for Number of Instances.

202) Click **OK** ✅ from the CirPattern PropertyManager. View the results in the PropertyManager.

In the next section, add an Appearance to the Wheel. Use the Appearances PropertyManager to apply colors, material appearances, and transparency to parts and assembly components. The Appearance tool does not modify the mass properties.

Activity: Add an Appearance to the Wheel.

Add an Appearance to the Wheel.
203) Right-click **Wheel** in the FeatureManager.

204) Click the **Appearances** drop-down arrow.

205) Click the **WHEEL** icon as illustrated. The color PropertyManager is displayed.

206) Click a **color (yellow)** from the Color dialog box.

207) Click **OK** ✔ from the color PropertyManager. View the results in the Graphics window.

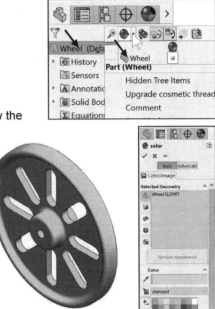

Display an Isometric view - Shaded With Edges.
208) Press the **space bar** to display the Orientation dialog box.

209) Click the **Isometric view** icon. You can also access the Isometric view tool from the Heads-up View toolbar.

210) Click the **Shaded With Edges** icon.

Save the Wheel.
211) Click **Save**.

In the next section, apply the Measure tool. The Measure tool provides the ability to measure distance, angle, and radius in sketches, 3D models, assemblies, or drawings. The tool also measures the size of and between lines, points, surfaces, and planes.

Activity: Apply the Measure tool.

Apply the Measure tool.
212) Click the **Evaluate** tab from the CommandManager.

213) Click the **Measure** tool. The Measure - Wheel dialog box is displayed.

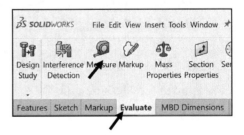

214) Click the **top face (Revolve1)** as illustrated. View the results. Explore the ability of the Measure -Wheel dialog box.

215) Click **Close** to the close the Measure - Wheel dialog box.

Save the Wheel.

216) Click **Save** 💾. You are finished with the part.

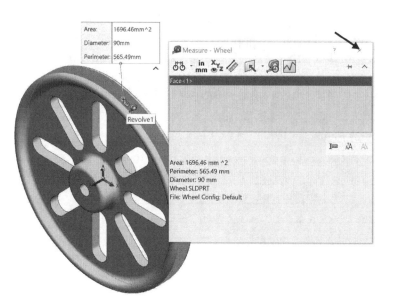

You created a new part (Wheel) that is used in the next chapter to create the Fly Wheel sub-assembly. The Fly Wheel sub-assembly with other provided components is used to create the final Stirling Engine assembly.

Summary

A SOLIDWORKS session was stated. You created a new part (Wheel) with user defined document and system properties. The correct Sketch plane and orientation for the Wheel was selected.

The Wheel part is used in the Fly Wheel sub-assembly. The sub-assembly is used in the final Stirling Engine assembly.

The following sketch tools were utilized: Circle, Line, Centerline, Center point Straight Slot and Mirror Entities.

The following features were utilized: Extruded Boss/Base, Extruded Cut, Revolved Boss/Boss, Hole Wizard, Circular Pattern and Fillet.

Material and Appearance were applied with the proper design intent.

You applied the Measure tool.

During the creation of the Wheel you learned about various Sketch states, editing Sketches and Features along with applying the following geometric relations: Coincident, Vertical, Horizontal, Tangent and Midpoint.

💡 If you modify a document property from an Overall drafting standard, a modify message is displayed as illustrated.

Overall drafting standard
ANSI-MODIFIED

Derived from: ANSI

The FeatureManager consists of various tabs:

- *FeatureManager design tree* tab.

- *PropertyManager* tab.

- *ConfigurationManager* tab.

- *DimXpertManager* tab.

- *DisplayManager* tab.

- *CAM FeatureManager tree* tab.

- *CAM Operation tree* tab.

- *CAM Tools tree* tab.

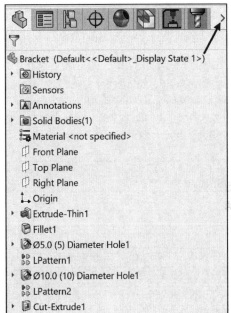

Click the ⬅️ direction arrows to expand or collapse the FeatureManager design tree.

CommandManager and FeatureManager tabs and tree folders will vary depending on system set-up and SOLIDWORKS Add-ins.

💡 Use the Search feature in the Matieral dialog box, to quickly located matieral type.

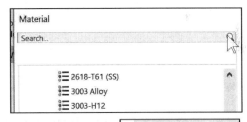

Download all model files from the SDC Publications website www.SDCpublications.com/downloads/978-1-63057-637-0.

Note: <u>Save to This PC</u>: Saves the document to your local hard drive in the folder which was last opened from. <u>Save to 3DEXPERIENCE</u>: Opens the Save as New dialog box to save the file to a Collaborative space on the **3D**EXPERIENCE platform. The Save to 3DEXPERIENCE option is displayed if you purchased 3DEXPERIENCE education roles and are logged into the **3D**EXPERIENCE platform. <u>Save As</u>: Saves the document to a new file name that becomes the active document without saving the original document. <u>Save All</u>: Saves all the open files that have been modified since they were last saved.

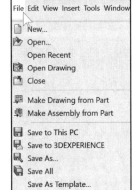

In Chapter 3, establish a SOLIDWORKS session and create two new assemblies with user defined document properties:

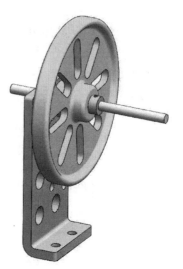

- Fly Wheel.

- Stirling Engine.

Insert the following Standard and Quick mate types: Coincident, Concentric, Distance and Tangent.

Utilize the following assembly tools: Insert Component, Suppress, Un-suppress, Mate, Move Component, Rotate Component, Interference Detection, Hide, Show, Flexible, Ridge, and Multiple mate mode.

Create an Exploded View with animation.

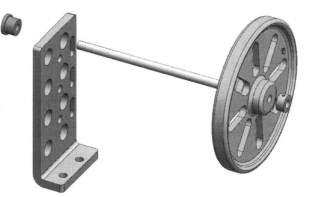

Apply the Measure and Mass Properties tool to modify a component in the Stirling Engine assembly.

Save the final Stirling Engine assembly, using the SOLIDWORKS Pack & Go tool.

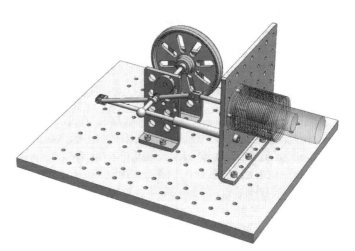

Questions

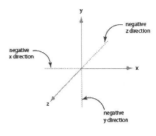

1. Describe the steps in starting a SOLIDWORKS session.

2. Describe the procedure to start a SOLIDWORKS 2D sketch.

3. Explain the steps required to modify units in a SOLIDWORKS part from inches to millimeters.

4. Identify the three default $\perp$ Reference planes in SOLIDWORKS.

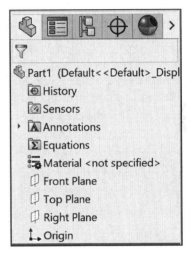

5. True or False. Add geometric relations, then dimensions in a sketch. This keeps the user from having too many unnecessary dimensions. This also helps to show the design intent of the model.

6. True or False. Geometric relations are constraints that control the size and position of the sketch entities.

7. True or False. Model about the Origin; this provides a point of reference for your dimensions to fully define the sketch.

8. True or False. Features are the building blocks of parts. The Extruded Boss/Base feature requires a Sketch plane, sketch and depth.

9. Describe an Extruded Cut feature in SOLIDWORKS.

10. The dimension sketch color black, indicates that the 2D sketch is _____ defined.

11. The dimension sketch color blue, indicates that the 2D sketch is _____ defined.

12. The dimension sketch color red, indicates that the 2D sketch is _____ defined.

13. Identify three Geometric sketch relations in SOLIDWORKS.

14. True or False. The plane you select for your first sketch determines the orientation of the part. Selecting the correct plane to start your model is very important.

Exercises

Exercise 2.1: Identify the Sketch plane for the Boss-Extrude1 (Base) feature as illustrated. Simplify the number of features.

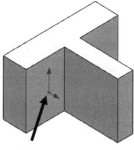

A: Top Plane

B: Front Plane

C: Right Plane

D: Left Plane

Correct answer _____.

Create the part. Dimensions are arbitrary.

Exercise 2.2: Identify the Sketch plane for the Boss-Extrude1 (Base) feature as illustrated. Simplify the number of features.

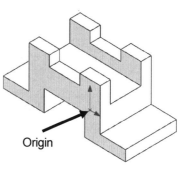

A: Top Plane

B: Front Plane

C: Right Plane

D: Left Plane

Correct answer _____.

Create the part. Dimensions are arbitrary.

Exercise 2.3: Identify the Sketch plane for the Boss-Extrude1 (Base) feature as illustrated. Simplify the number of features.

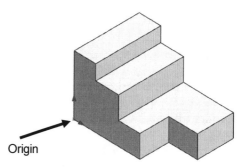

A: Top Plane

B: Front Plane

C: Right Plane

D: Left Plane

Correct answer _____.

Create the part. Dimensions are arbitrary.

Exercise 2.4: FLATBAR - 3HOLE Part

Create an ANSI, IPS FLATBAR - 3HOLE part document.

- Utilize the Front Plane for the Sketch plane. Insert an Extruded Base (Boss-Extrude1) feature. Do not display Tangent Edges.

- Create an Extruded Cut feature. This is your seed feature. Apply the Linear Pattern feature. The FLATBAR - 3HOLE part is manufactured from 0.06in., [1.5mm] 6061 Alloy.

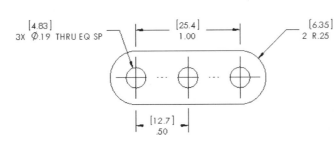

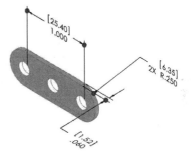

Exercise 2.5: FLATBAR - 5HOLE Part

Create an ANSI, IPS, FLATBAR - 5HOLE part as illustrated.

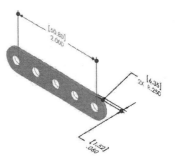

- Utilize the Front Plane for the Sketch plane. Insert an Extruded Base (Boss-Extrude1) feature.

- Create an Extruded Cut feature. This is your seed feature. Apply the Linear Pattern feature. The FLATBAR - 5HOLE part is manufactured from 0.06in, [1.5mm] 6061 Alloy.

- Calculate the required dimensions for the FLATBAR - 5HOLE part. Use the following information: Holes are .500in. on center, Radius is .250in., and hole diameter is .190in.

Think design intent. When do you use the various End Conditions and Geometric sketch relations? What are you trying to do with the design? How does the component fit into an assembly?

Exercise 2.6: Simple Block Part

Create the illustrated ANSI part. Note the location of the Origin in the illustration.

- Calculate the overall mass of the illustrated model.

- Apply the Mass Properties tool.

- Think about the steps that you would use to build the model.

- Review the provided information carefully.

- Units are represented in the IPS (inch, pound, second) system.

- A = 3.50in, B = .70in

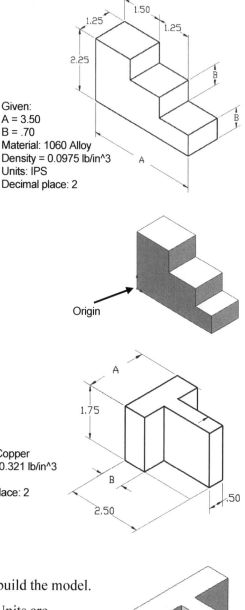

Given:
A = 3.50
B = .70
Material: 1060 Alloy
Density = 0.0975 lb/in^3
Units: IPS
Decimal place: 2

Origin

Exercise 2.7: Simple Block Part

Create the illustrated ANSI part. Note the location of the Origin in the illustration.

Create the sketch symmetric about the Front Plane. The Front Plane in this problem is **not** your Sketch Plane. Utilize the Blind End Condition in Direction 1.

Given:
A = 3.00
B = .75
Material: Copper
Density = 0.321 lb/in^3
Units: IPS
Decimal place: 2

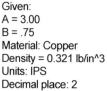

- Calculate the overall mass of the illustrated model.

- Apply the Mass Properties tool.

- Think about the steps that you would use to build the model.

- Review the provided information carefully. Units are represented in the IPS (inch, pound, second) system.

- A = 3.00in, B = .75in

Note: Sketch1 is symmetrical.

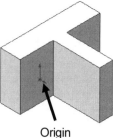

Origin

Exercise 2.8: Simple Block Part

Create an ANSI part from the illustrated model. Note the location of the Origin in the illustration.

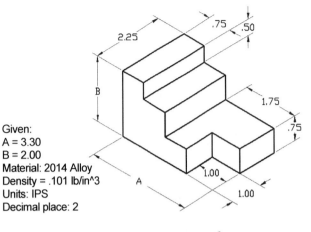

- Calculate the volume of the part and locate the Center of mass with the provided information.

Given:
A = 3.30
B = 2.00
Material: 2014 Alloy
Density = .101 lb/in^3
Units: IPS
Decimal place: 2

- Apply the Mass Properties tool.

- Think about the steps that you would use to build the model.

- Review the provided information carefully.

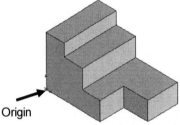

Origin

Exercise 2.9: Simple Block Part

Create an ANSI, MMGS part from the illustrated drawing: Front, Top, Right and Isometric views.

Note the location of the Origin in the illustration. The drawing views are displayed in Third Angle Projection.

- Apply 1060 Alloy for material.

- Calculate the Volume of the part.

- Locate the Center of mass.

Think about the steps that you would use to build the model. The part is symmetric about the Front Plane.

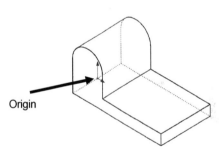

Origin

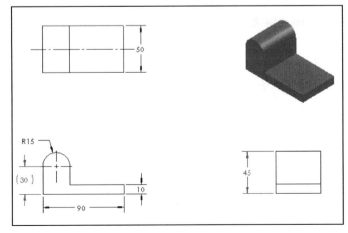

Exercise 2.10: Simple Block Part

Create the ANSI, MMGS part from the illustrated drawing: Front, Top, Right and Isometric views.

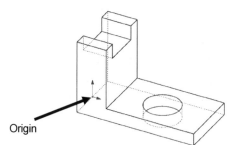

Origin

- Apply 1060 Alloy for material.

- The part is symmetric about the Front Plane.

- Calculate the Volume of the part and locate the Center of mass.

Think about the steps that you would use to build the model.

The drawing views are displayed in Third Angle Projection.

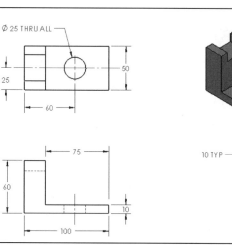

Exercise 2.11: LINKAGE-2 Assembly

Create the LINKAGE-2 assembly.

- Copy the LINKAGE assembly from the Chapter 2 Homework folder to your local hard drive.

- Open the assembly.

- Select Save As from the drop-down Menu bar.

- Check the Save as copy and open check box.

- Enter LINKAGE-2 for file name. LINKAGE-2 ASSEMBLY for description.

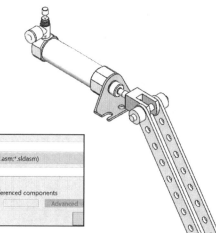

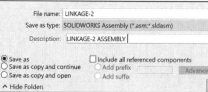

The FLATBAR-3HOLE part was created in Exercise 2.4. Utilize two AXLE parts, four SHAFT COLLAR parts, and two FLATBAR-3HOLE parts to create the LINKAGE-2 assembly as illustrated.

The AXLE, SHAFT COLLAR and FLATBAR-3HOLE is provided in the Homework folder.

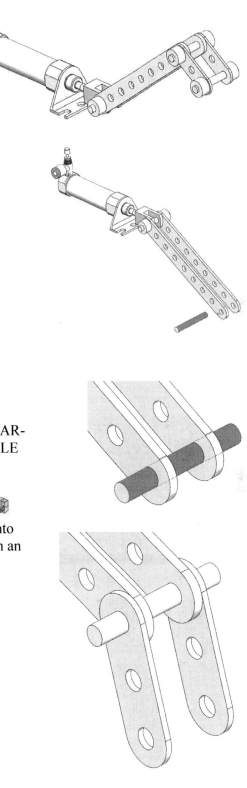

- Insert the first AXLE part.

- Insert a Concentric mate.

- Insert a Coincident mate.

- Insert the first FLATBAR-3HOLE part.

- Insert a Concentric mate.

- Insert a Coincident mate.

- Perform the same procedure for the second FLATBAR-3HOLE part.

- Insert a Parallel mate between the 2 FLATBAR-3HOLE parts. Note: The 2 FLATBAR-3HOLE parts move together.

☀ When a component is in the Lightweight state, only a subset of its model data is loaded into memory. The remaining model data is loaded on an as-needed basis.

☀ When a component is fully resolved, all its model data is loaded into memory.

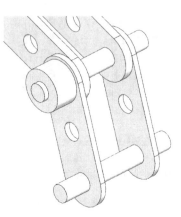

- Insert the second AXLE part.

- Insert a Concentric mate.

- Insert a Coincident mate.

- Insert the first SHAFT-COLLAR part.

- Insert a Concentric mate.

- Insert a Coincident mate.

- Perform the same tasks to insert the other three required SHAFT-COLLAR parts as illustrated.

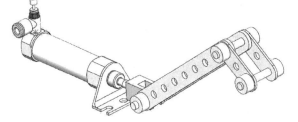

Exercise 2.12: ROCKER Assembly

Create a ROCKER assembly. The ROCKER assembly consists of two AXLE parts, two FLATBAR-5HOLE parts, and two FLATBAR-3HOLE parts. **Note**: The parts are in the Homework Chapter folder.

The FLATBAR-3HOLE parts are linked together with the FLATBAR-5HOLE.

The three parts rotate clockwise and counterclockwise, above the Top Plane. Create the ROCKER assembly.

- Insert the first FLATBAR-5HOLE part. The FLATBAR-5HOLE is fixed to the Origin of the ROCKER assembly.

- Insert the first AXLE part.

- Insert a Concentric mate.

- Insert a Coincident mate.

- Insert the second AXLE part.

- Insert a Concentric mate.

- Insert a Coincident mate.

- Insert the first FLATBAR-3HOLE part.

- Insert a Concentric mate.

- Insert a Coincident mate.

- Insert the second FLATBAR-3HOLE part.

- Insert a Concentric mate.

- Insert a Coincident mate.

- Insert the second FLATBAR-5HOLE part.

- Insert the required mates.

Note: The end holes of the second FLATBAR-5HOLE are concentric with the end holes of the FLATBAR-3HOLE parts.

Note: In mechanical design, the ROCKER assembly is classified as a mechanism. A Four-Bar Linkage is a common mechanism comprised of four links.

Link1 is called the Frame.

The AXLE part is Link1.

Link2 and Link4 are called the Cranks.

The FLATBAR-3HOLE parts are Link2 and Link4. Link3 is called the Coupler. The FLATBAR-5HOLE part is Link3.

If an assembly or component is loaded in a Lightweight state, right-click the assembly name or component name from the FeatureManager. Click Set Lightweight to Resolved.

Determine the static and dynamic behavior of mates in each sub-assembly before creating the top level assembly.

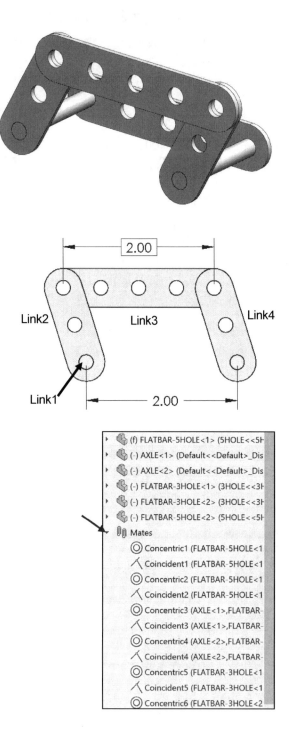

Exercise 2.13: 4 Bar linkage

Create the 4 bar linkage assembly as illustrated. Create the five components. Assume dimensions. **Note**: View the MP4 file in the Chapter Homework folder.

In an assembly, fix (f) the first component to the origin or fully define it to the three default reference planes.

Insert all needed mates to simulate the movement of a 4 bar linkage assembly.

Read the section on Coincident, Concentric and Distance mates in SOLIDWORKS help.

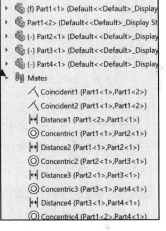

Create a base with text for extra credit.

Below are a few sample models from my Freshman Engineering class.

Note the different designs to maintain the proper movement of a 4 bar linkage using a base.

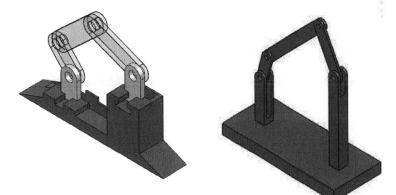

Exercise 2.14: Fill Pattern

Create a Polygon Layout Fill Pattern feature. Apply the seed cut option.

1. Open **Fill Pattern 2.14** from the Chapter 2 Homework folder.

2. Click the **Fill Pattern** 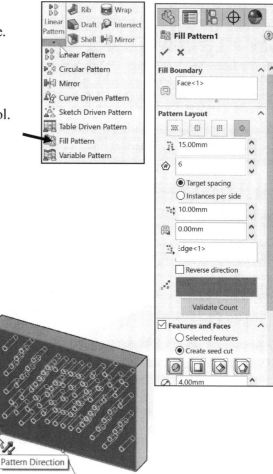 Features tool. The Fill Pattern PropertyManager is displayed.

3. Click the **Front face** of Extrude1. Face<1> is displayed in the Fill Boundary box. The direction arrow points to the right.

4. Click **Polygon** for Pattern Layout.

5. Click **Target spacing**.

6. Enter **15**mm for Loop Spacing.

7. Enter **6** for Polygon sides.

8. Enter **10**mm for Instances Spacing.

9. Enter **10**mm for Margins. View the direction arrow.

10. Click **Create seed cut**.

11. Click **Circle** for Features to Pattern.

12. Enter **4mm** for Diameter.

13. Click **OK** ✔ from the Fill Pattern PropertyManager. Fill Pattern1 is created and is displayed in the FeatureManager.

14. **View** the results.

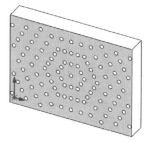

Exercise 2.15: Fill Pattern

Create a Circular Layout Fill Pattern feature.

1. **Copy** the Chapter 2 Homework folder to your hard drive

2. Open **Fill Pattern 2.15** from the Chapter 2 Homework folder.

3. Click the **Fill Pattern** Features tool. The Fill Pattern PropertyManager is displayed.

4. Click the **top face** of Boss-Extrude1. Face<1> is displayed in the Fill Boundary box. The direction arrow points to the back. Click **Circular** for Pattern Layout.

5. Enter **.10**in for Loop Spacing.

6. Click **Target spacing**.

7. Enter .**10**in Instance Spacing.

8. Enter **.05**in for Margins. Edge<1> is selected for Pattern Direction.

9. Click **inside** the Features to Pattern box.

10. Click **Boss-Extrude2** from the fly-out FeatureManager. Boss-Extrude2 is displayed in the Features to Pattern box.

11. Click **OK** from the Fill Pattern PropertyManager. Fill Pattern1 is created and is displayed in the FeatureManager.

12. **View** the results.

Exercise 2.16: Tangent Mate

Create a Tangent mate between the roller-wheel and the cam on both sides to simulate movement per the avi file.

1. **Copy** the Chapter 2 Homework folder to your hard drive.

2. Open **Tangent** from the Chapter 2 Homework\Tangent folder. The assembly is displayed.

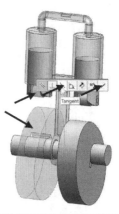

Create a Tangent mate between the roller-wheel and the cam on both sides. The cam was created with imported geometry (curve).

A Tangent mate places the selected items tangent to each other. At least one selected item must be either a conical, cylindrical, spherical face. An Angle mate places the selected items at the specified angle to each other.

3. Click the **Mate** tool from the Assembly tab.

4. **Pin** the Mate PropertyManager. Click the **Standard** tab.

Create the first Tangent mate.

5. Click the **contact face** of the left cam as illustrated.

6. Click the **contact face** of the left roller-wheel as illustrated.

7. Click **Tangent** mate from the Pop-up menu.

8. Click **Add/Finish mate**. Tangent1 is created.

Create the second Tangent mate.

9. Click the **contact face** of the right cam.

10. Click the **contact face** of the right roller-wheel.

11. Click **Tangent** mate from the PropertyManager.

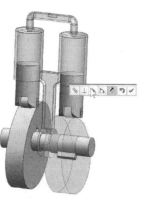

12. Click Add/Finish mate from the Pop-up menu. Tangent2 is created.

13. **Un-Pin** the Mate PropertyManager.

14. Click **OK** from the Mate PropertyManager.

15. Display an **Isometric view**.

16. **Move** the cam. View the results.

Chapter 3

Assembly Modeling - Bottom up method

Below are the desired outcomes and usage competencies based on the completion of Chapter 3.

Desired Outcomes:	Usage Competencies:

• Assembly documents:
 	
   • Set drafting standard.
 	
   • Set units and precision.
 	
• Sub-assembly:
 	
   • Fly Wheel.
 	
• Final assembly:
 	
   • Stirling Engine.
 	
• Exploded Sub-assembly.
 	
• Save An Assembly Document.
	• Set Document Properties as they relate to an Assembly.
	
	• Comprehend the assembly process. Insert parts/sub-components into an assembly.
	
	• Ability to apply Standard mates and Quick mates (Coincident, Concentric, Distance and Tangent) in an assembly using the Bottom-up design approach.
	
	• Utilize the following assembly tools: Insert Component, Mate, Hide, Show, Move, Rotate, Modify, Flexible, Ridge and Multiple mate mode.
	
	• Apply the Measure and Mass Properties tool.
	
	• Incorporate design changes into an assembly.

Notes:

Chapter 3 - Assembly Modeling - Bottom up method

Chapter Objective

Create two assemblies. The first assembly is the Fly Wheel. The second assembly is the Stirling Engine. The Fly Wheel is a sub-assembly of the final Stirling Engine assembly.

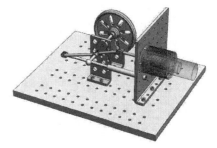

On the completion of this chapter, you will be able to:

- Create a new assembly document.

- Set user defined document properties.

- Understand assembly modeling approach.

- Comprehend linear and rotational motion.

- Insert parts and sub-assemblies in an assembly.

- Insert and edit the following Standard mates:

  - Coincident, Concentric, Distance and Tangent.

- Incorporate design changes in an assembly.

- Edit and Modify features and components in an assembly.

- Utilize the following assembly tools: Insert Component, Mate, Hide, Show, Move, Rotate, Modify, Flexible, Ridge and Multiple mate mode.

- Create an Exploded View with animation.

- Apply the Measure and Mass Properties tool.

- Apply the Pack and Go tool.

Activity: Start a SOLIDWORKS Session. Create a New Assembly.

Start a SOLIDWORKS session.

1) Double-click the **SOLIDWORKS 2024 icon** from the desktop.

2) **Close** the Welcome - SOLIDWORKS dialog box.

Create a New Assembly Document.

3) Click **New** ☐ from the Menu bar or click **File**, **New** from the Menu bar menu. The New SOLIDWORKS Document dialog box is displayed.

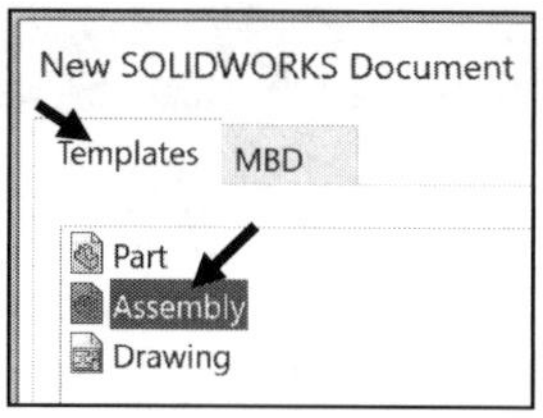

4)   Click the **Templates** tab.

5)   Double-click **Assembly** from the New SOLIDWORKS Document
        dialog box. The Begin Assembly PropertyManager and the
        Window Open dialog box is displayed.

6)   Click **Cancel** in the Open dialog box.

7)   Click **Cancel** ✖ from the Assembly FeatureManager. Assem1 is
        displayed in the FeatureManager.

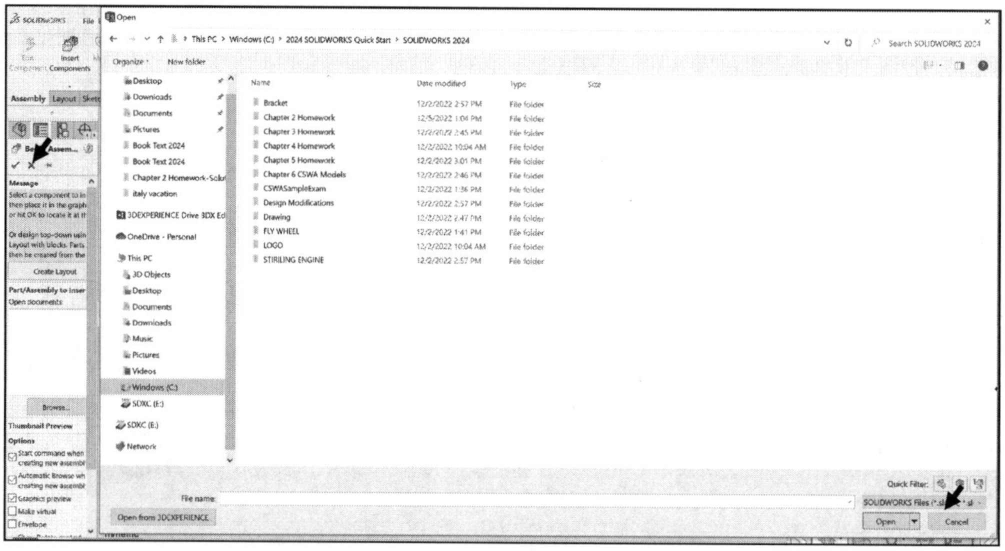

The first system default Assembly filename is Assem1. The system attaches the
.sldasm to the created assembly. The second created assembly in the same session
increments to the filename Assem2.

The Begin Assembly
PropertyManager and the
Insert Component
PropertyManager is
displayed when a new or
existing assembly is opened,
if the Start command when
creating new assembly box
is checked.

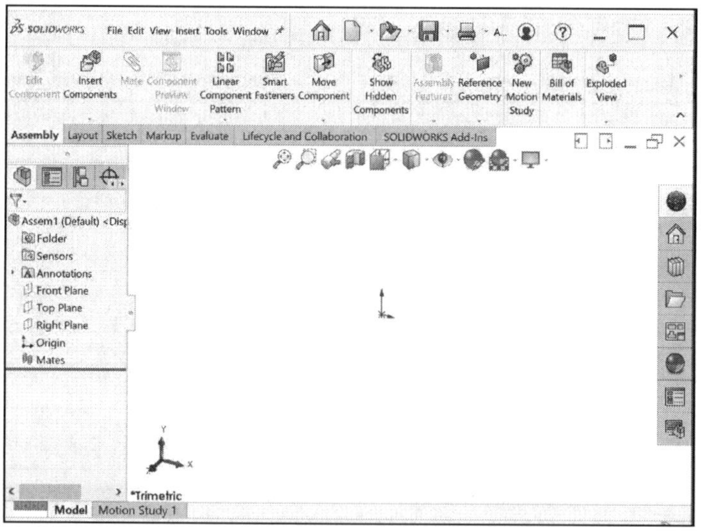

Display the origin.

8)   Click **View**, **Hide/Show**,
        Origins from the Main
        menu. The origin is
        displayed in the
        Graphics window.

Directional input refers by default to the global coordinate system (X, Y and Z), which is based on Plane1 with its origin located at the origin of the part or assembly. Plane1 (Front) is the first plane that appears in the FeatureManager design tree and can have a different name. The reference triad shows the global X-, Y- and Z-directions.

Activity: Set Document Properties for the Fly Wheel Assembly.

---

9)   Click **Options** ⚙ ˙ from the Menu bar. The System Options General dialog box is displayed.

10)   Click the **Document Properties** tab.

11)   Select **ANSI** from the Overall drafting standard drop-down menu. Various Detailing options are available depending on the selected standard.

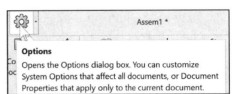

💡 Various detailing options are available depending on the selected standard.

The Overall drafting standard determines the display of dimension text, arrows, symbols, and spacing. Units are the measurement of physical quantities. Millimeter dimensioning and decimal inch dimensioning are the two most common unit types specified for engineering parts and drawings.

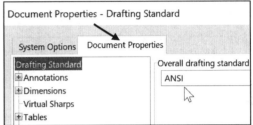

Set Document Properties. Enter units and precision.

12)   Click the **Units** folder.

13)   Click **MMGS** (millimeter, gram, second) for Unit system.

14)   Select **.12**, (two decimal places) for Length basic units.

15)   Click **OK** from the Document Properties - Units dialog box. The Assembly FeatureManager is displayed.

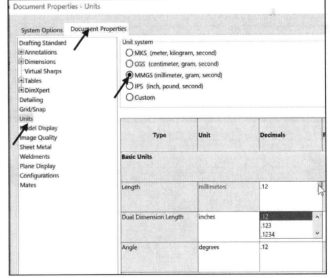

💡 The Origin ↳ represents the intersection of the Front, Top and Right Planes.

Assembly Modeling Approach

In SOLIDWORKS, components and their assemblies are directly related through a common file structure. Changes in the components directly affect the assembly and vice versa. You can create assemblies using the Bottom-up assembly approach, Top-down assembly approach or a combination of both methods. This chapter focuses on the Bottom-up assembly approach.

The Bottom-up approach is the traditional method that combines individual components. Based on design criteria, the components are developed independently. The three major steps in a Bottom-up assembly approach are create each component independent of any other component in the assembly, insert the components into the assembly, and mate the components in the assembly as they relate to the physical constraints of your design.

In the Top-down assembly approach, major design requirements are translated into layout sketches, assemblies, sub-assemblies and components.

In the Top-down approach, you do not need all of the required component design details. Individual relationships are required.

Example: A computer. The inside of a computer can be divided into individual key sub-assemblies such as a motherboard, hard drive, power supply, etc. Relationships between these sub-assemblies must be maintained for proper fit.

Linear Motion and Rotational Motion

In dynamics, motion of an object is described in linear and rotational terms. Components possess linear motion along the x, y and z-axes and rotational motion around the x, y and z-axes. In an assembly, each component has *six degrees* of freedom: three translational (linear) and three rotational. Mates remove degrees of freedom. All components are rigid bodies. The components do not flex or deform.

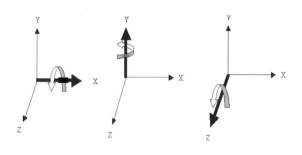

Activity: Insert the First Component into the Fly Wheel Assembly.

The first component is the foundation of the assembly. The Bracket is the first component in the Fly Wheel assembly. The Bushing is the second component in the Fly Wheel assembly. All needed components for this chapter are located in the SOLIDWORKS 2024\FLY WHEEL folder.

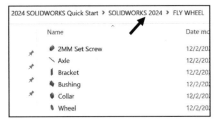

Add components to assemblies utilizing the following techniques:

- Utilize the Insert Components Assembly tool.

- Utilize Insert, Component from the Menu bar.

- Drag a component from Windows Explorer into the Assembly.

- Drag a component from the SOLIDWORKS Design Library into the Assembly.

- Drag a component from an Open part file into the Assembly.

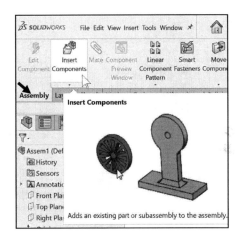

Insert the Bracket component.

16) Click the **Assembly** tab from the CommandManager.

17) Click **Insert Components** from the Assembly toolbar. The Insert Component PropertyManager and the Window Open dialog box is displayed.

18) Double-click the **Bracket** part from the SOLIDWORKS 2024\FLY WHEEL folder.

19) Click **OK** from the Insert Component PropertyManager. The Bracket is <u>fixed</u> to the origin. It cannot translate or rotate.

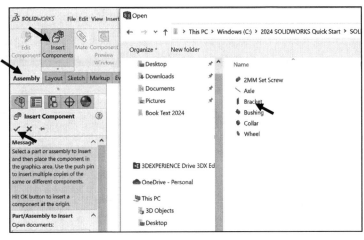

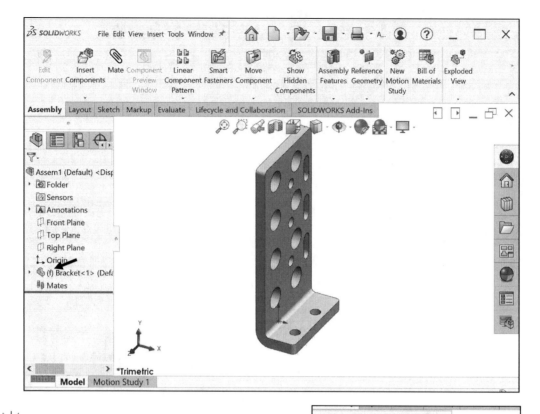

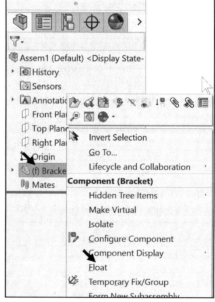

💡 To fix the first component to the Origin, click OK ✔ from the Begin Assembly PropertyManager or click the Origin in the Graphics window.

💡 To remove the fixed state (f), right-click the **fixed component name** in the FeatureManager. Click **Float**. The component is free to move.

Display an Isometric view of the model. Press the **space bar** to display the Orientation dialog box. Click the **Isometric view** icon.

Deactive the Origin. Save the Assembly.

20) Click **View**, **Hide/Show** from the Main menu. Click the **Origins** button.

21) Click **Save As** from the Drop-down menu bar.

22) Browse to the **SOLIDWORKS 2024** folder.

23) Enter **Fly Wheel** for the File name.

24) Enter **Fly Wheel** for the Description.

25) Click **Save** from the Save As dialog box. The Fly Wheel Assembly FeatureManager is displayed.

Mate Types

Mates provide the ability to create geometric relationships between assembly components. Mates define the allowable directions of rotational or linear motion of the components in the assembly. Move a component within its degrees of freedom in the Graphics window to view the behavior of an assembly.

Standard Mates:

The Standard mates tab is the default tab. The mate tabs provide the ability to insert a *Standard*, *Advanced, Mechanical* mate or to address *Analysis*. Standard mate types:

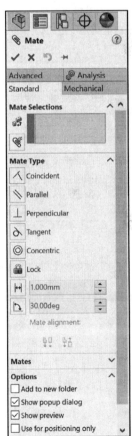

- **Coincident**: Locate the selected faces, edges, points or planes so they use the same infinite line. A Coincident mate positions two vertices for contact.

- **Parallel**: Locate the selected items to lie in the same direction and to remain a constant distance apart. A parallel mate permits only translational motion of a single part with respect to another. No rotation is allowed.

- **Perpendicular**: Locate the selected items at a 90-degree angle to each other. The perpendicular mate allows both translational and rotational motion of one part with respect to another.

- **Tangent**: Locates the selected items in a tangent mate. At least one selected item must be either a conical, cylindrical, or spherical face.

- **Concentric**: Locates the selected items so they can share the same center point. You can prevent the rotation of components that are mated with concentric mates by selecting the Lock Rotation option. Locked concentric mates are indicated by an icon in the FeatureManager design tree.

- **Lock**: Maintains the position and orientation between two components.

- **Distance**: Locates the selected items with a specified distance between them. Use the drop-down arrow box or enter the distance value directly.

- **Angle**: Locates the selected items at the specified angle to each other. Use the drop-down arrow box or enter the angle value directly.

- There are two Mate Alignment options. The **Aligned** option positions the components so that the normal vectors from the selected faces point in the same direction. The **Anti-Aligned** option positions the components so that the normal vectors from the selected faces point in opposite directions.

Advanced Mates:

The Advanced Mate types:

- **Profile Center**: Mate to center automatically center-aligns common component types such as rectangular and circular profiles to each other.

- **Symmetric**: Force two similar entities to be symmetric about a planar face or plane.

- **Width**: Center a tab within the width of a groove. The Width mate is used to replace the Symmetric mate where components have tolerance and a gap rather than a tight fit.

- **Path**: Enable any point on a component to be set to follow a defined path.

- **Linear/Linear Coupler**: Establish a relationship between the translation of one component and the translation of another component.

- **Limit**: Allow components to move within a range of values for distance and angle. Select the angle and distance from the provided boxes. Specify a starting distance or angle as well as a maximum and minimum value.

- **Distance**: Locate the selected items with a specified distance between them. Use the drop-down arrow box or enter the distance value directly (Limit mate).

- **Angle**: Locate the selected items at the specified angle to each other. Use the drop-down arrow box or enter the angle value directly.

- **Mate alignment**: Toggle the mate alignment as necessary. The options are:
  - **Aligned**: Locate the components so the normal or axis vectors for the selected faces point in the same direction.
  - **Anti-Aligned**: Locate the components so the normal or axis vectors for the selected faces point in the opposite direction.

Mechanical Mates:

The Mechanical Mate types:

- **Cam**: Force a plane, cylinder, or point to be tangent or coincident to a series of tangent extruded faces. Four conditions can exist with the Cam mate: *Coincident, Tangent, CamMateCoincident* and *CamMateTangent*. Create the profile of the cam from lines, arcs and splines as long as they are tangent and form a closed loop.

- **Slot**: Constrain the movement of a bolt or a slot to within a slot hole. You can mate bolts to straight or arced slots and you can mate slots to slots. You can select an axis, cylindrical face, or a slot to create slot mates.

- **Hinge**: Limit the movement between two components to one rotational degree of freedom. It has the same effect as adding a Concentric mate plus a Coincident mate. You can also limit the angular movement between the two components.

- **Gear**: Force two components to rotate relative to one another around selected axes. The Gear mate provides the ability to establish gear type relations between components without making the components physically mesh. Note: SOLIDWORKS provides the ability to modify the gear ratio without changing the size of the gears. Align the components before adding the Mechanical gear mate.

- **Rack and Pinion**: Linear translation of one component (the rack) causes circular rotation in another component (the pinion), and vice versa. You can mate any two components to have this type of movement relative to each other. The components do not need to have gear teeth.

- **Screw**: Constrain two components to be concentric, and adds a pitch relationship between the rotation of one component and the translation of the other. Translation of one component along the axis causes rotation of the other component according to the pitch relationship. Likewise, rotation of one component causes translation of the other component. Align the components before adding the Mechanical screw mate.

- **Universal Joint**: Permit the transfer of rotations from one rigid body to another. This mate is particularly useful to transfer rotational motion around corners, or to transfer rotational motion between two connected shafts that are permitted to bend at the connection (drive shaft on an automobile).

Mates reflect the physical behavior of a component in an assembly. In this project, the two most common Mate types are Concentric and Coincident.

Download all model files from the SDC Publications website www.SDCpublications.com/downloads/978-1-63057-637-0 to a local hard drive. All components are provided.

Activity: Insert and Mate the Second Component - Bushing.

Insert the Bushing part into the Fly Wheel Assembly.

26)   Click the **Assembly** tab from the CommandManager.

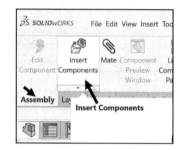

27)   Click **Insert Component** from the Assembly toolbar. The Insert Component PropertyManager and the Window Open dialog box is displayed.

28)   Double-click the **Bushing** part from the SOLIDWORKS 2024\FLY WHEEL folder.

29)   Click a position to the **left** of the Bracket. The Bushing is displayed.

Insert a Concentric Mate between the outside cylindrical face of the Bushing and the cylindrical face of the top right hole.

30)   Click the **outside cylindrical face** of the Bushing.

31)   Hold the **Ctrl** key down.

32)   Click **cylindrical face** of the top right hole as illustrated.

33)   Release the **Ctrl** key. The Mate Pop-up menu is displayed.

34)   Click **Concentric** ◎ from The Mate Pop-up menu. A Concentric mate

locates the selected items to share the same center point. **Note: The Pop-up menu to Lock Rotation or to Flip Mate Aligment icons.**

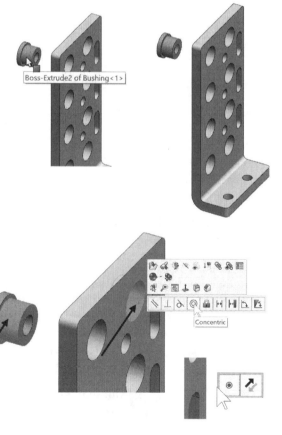

Insert a Coincident Mate between the flat circular face of the Bushing and the back face of the Bracket.

35) **Rotate** the model to view the back face of the Bracket.

36) **Click and drag** the Bushing away from the back face of the Bracket.

37) Click the **flat circular face** of the Bushing.

38) **Rotate** the model to view the flat circular face of the Bracket.

39) Hold the **Ctrl** key down.

40) Click the **flat face** of the Bracket.

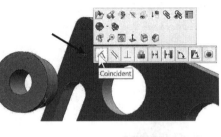

41) Release the **Ctrl** key. The Mate Pop-up menu is displayed.

42) Click **Coincident** ⟨ from The Mate Pop-up menu. A Coincident mate locates the selected faces, edges or planes so they use the same infinite lines. **Note: The Pop-up menu Flip Mate Aligment icon.**

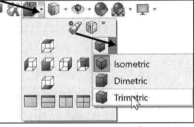

Display a Trimetric view.

43) Click **Trimetric view** ▢ from the Heads-up toolbar.

Expand the Mates folder.

44) **Expand** the Mates folder as illustrated. View the results.

Save the model.

45) Click **Save** 💾. Note, Tangent edges are displayed on the model. To remove Tangent Edges, click Options, System Options, Display, Removed.

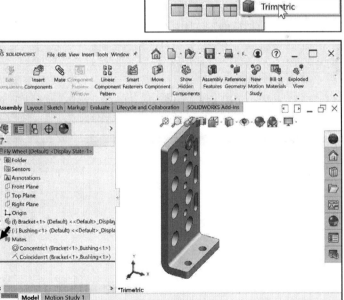

💡 If you delete a Mate and then recreate it, the Mate numbers will be different (increase).

💡 Determine the static and dynamic behavior of mates in each sub-assembly before creating the top-level assembly.

Activity: Insert and Mate the Third Component - Axle.

Insert the Axle part into the Fly Wheel Assembly.

46) Click the **Assembly** tab in the CommandManager.

47) Click **Insert Component**  from the Assembly toolbar. The Insert Component PropertyManager and the Window Open dialog box is displayed.

48) Double-click the **Axle** part from the SOLIDWORKS 2024\FLY WHEEL folder.

49) Click a position to the **right** of the Bracket as illustrated. The Axle is displayed.

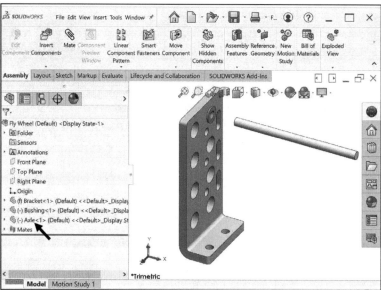

Insert a Concentric Mate between the circular face of the Axle and the circular face of the Bushing hole.

50) **Zoom-in** on the Axle and Bushing in the Graphics window.

51) Click the **cylindrical face** of the Axle.

52) Hold the **Ctrl** key down.

53) Click the inside **cylindrical face** of the Bushing hole.

54) Release the **Ctrl** key. The Mate Pop-up menu is displayed.

55) Click **Concentric** ◎ from The Mate Pop-up menu. Concentric2 is created.

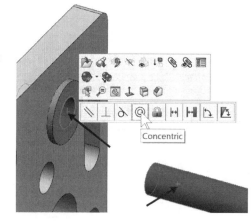

Create a Distance Mate (20.00mm) between the back face of the Bushing and the front face of the Axle.

56) **Rotate** the model to view the back face of the Bracket.

57) Click the **back face** of the Bushing.

58) Hold the **Ctrl** key down.

59) Click the **front face** of the Axle as illustrated.

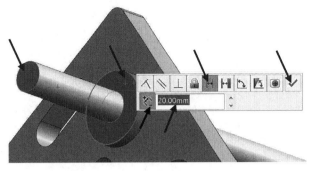

60) Release the **Ctrl** key. The Mate pop-up menu is displayed.

61) Click **Distance** ⊢⊣ mate from the Mate pop-up menu.

62) Enter **20**mm for Distance. **Note:** Click the Flip Direction icon if needed.

63) Click the **Green Check mark** ✔ from the Mate pop-up menu. Distance1 is created. A Distance mate locates the selected items with a specified distance between them.

Display a Trimetric view.

64) Click **Trimetric view** 📦 from the Heads-up View toolbar.

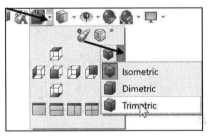

Save the model.

65) Click **Save** 💾.

Activity: Insert and Mate the Fourth Component - Wheel.

Insert the Wheel part into the Fly Wheel Assembly. The Wheel part was created in Chapter 2. If you did not create the part, insert the Wheel part from the SOLIDWORKS 2024\FLY WHEEL folder.

66) Click **Insert Component** 📦 from the Assembly toolbar. The Insert Component PropertyManager and Window Open dialog box is displayed.

67) Double-click the **Wheel** part from the SOLIDWORKS 2024\FLY WHEEL folder.

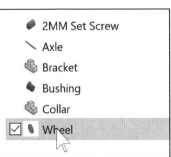

68) Click a **position** as illustrated in the Graphics window.

Insert a Concentric Mate between the inside cylindrical face of the hub and the outside cylindrical face of the Axle.

69) **Zoom-in** on the Wheel and Axle in the Graphics window.

70) Click the **inside cylindrical face** of the Wheel hole (Hub).

71) Hold the **Ctrl** key down.

72) Click the **cylindrical face** of the Axle.

73) Release the **Ctrl** key. The Mate Pop-up menu is displayed.

74) Click **Concentric** ◎ from the Mate Pop-up menu.

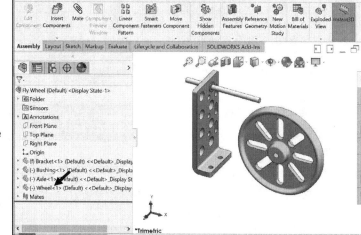

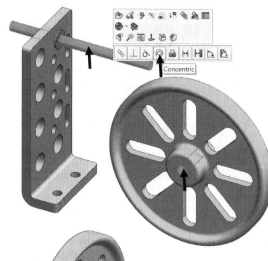

Insert a .1mm Distance Mate between the back face of the hub and the front face of the Bushing.

75) **Rotate** the assembly to view the back face of the hub.

76) Click the **back face** of the hub as illustrated.

77) **Rotate** and **Zoom in** on the front face of the Bushing.

78) Hold the **Ctrl** key down.

79) Click the **front face** of the Bushing.

80) Release the **Ctrl** key. The Mate Pop-up menu is displayed.

81) Click **Distance** ⊢⊣ mate from the Mate pop-up menu.

82) Enter **.1**mm for Distance.

83) Click the **Green Check mark** ✔ from the Mate pop-up menu. Distance2 is created. A Distance mate locates the selected items with a specified distance between them.

Fit the model to the Graphics window.
84) Press the **f** key.

Display a Trimetric view.
85) Click **Trimetric view** 🔲 from the Heads-up View toolbar.

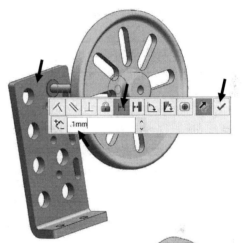

Save the model.
86) Click **Save** 💾 .

Rotate the Wheel in the Assembly.
87) Click on the **Wheel** and drag to view the Wheel rotate.

View the Created Mates.
88) **Expand** the Mates folder. View the created mates.

> ▸ 🧊 (f) Bracket<1> (Default<<Default>_Display
> ▸ 🧊 (-) Bushing<1> (Default<<Default>_Displa
> ▸ 🧊 (-) Axle<1> (Default<<Default>_Display S
> ▸ 🧊 (-) Wheel<1> (Default<<Default>_Display
> ▾ 📎 Mates
>   ◎ Concentric1 (Bracket<1>,Bushing<1>)
>   ⊼ Coincident1 (Bracket<1>,Bushing<1>)
>   ◎ Concentric2 (Bushing<1>,Axle<1>)
>   ⊢⊣ Distance1 (Axle<1>,Bushing<1>)
>   ◎ Concentric3 (Axle<1>,Wheel<1>)
>   ⊢⊣ Distance2 (Wheel<1>,Bushing<1>)

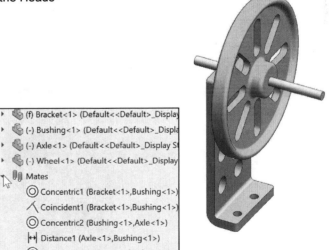

Activity: Insert and Mate the Fifth Component - Collar.

Insert the Collar part into the Fly Wheel Assembly.

89) Click **Insert Component** from the Assembly toolbar. The Insert Component PropertyManager and the Window Open dialog box is displayed.

90) Double-click the **Collar** part from the SOLIDWORKS 2024\FLY WHEEL folder.

91) Click a position to the **right** of the assembly as illustrated.

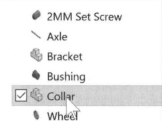

Insert a Concentric Mate between the inside cylindrical face of the Collar and the outside cylindrical face of the Axle.

92) **Zoom-in** on the Fly Wheel assembly in the Graphics window.

93) Click the **inside cylindrical face** of the Collar.

94) Hold the **Ctrl** key down.

95) Click the **outside cylindrical face** of the Axle.

96) Release the **Ctrl** key. The Mate Pop-up menu is displayed.

97) Click **Concentric** ⊚ from the Mate Pop-up menu.

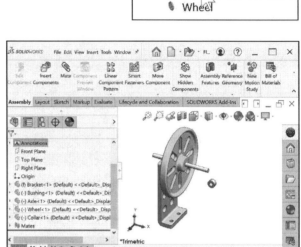

Insert Coincident Mate between the back face of the Collar and the front face of the hub.

98) **Rotate** the assembly to view the back face of the Collar.

99) Click the **back face** of the Collar as illustrated.

100) **Rotate** and **Zoom in** on the front face of the hub.

101) Hold the **Ctrl** key down.

102) Click the **front face** of the hub as illustrated.

103) Release the **Ctrl** key. The Mate Pop-up menu is displayed.

104) Click **Coincident** ⟋ from the Mate Pop-up menu.

Fit the model to the Graphics window.
105) Press the **f** key.

Display a Trimetric view.

106) Click **Trimetric view** ⬢ from the Heads-up View toolbar.

Rotate the Collar in the Assembly.
107) Click on the **Collar** and drag. The Collar is free to rotate.

108) Position the **2MM hole** towards the top as illustrated.

View the Created Mates.
109) **Expand** the Mates folder. View the created mates.

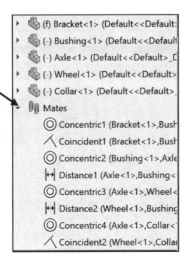

▸ 🧩 (f) Bracket<1> (Default<<Default:
▸ 🧩 (-) Bushing<1> (Default<<Default
▸ 🧩 (-) Axle<1> (Default<<Default>_D
▸ 🧩 (-) Wheel<1> (Default<<Default>
▸ 🧩 (-) Collar<1> (Default<<Default>_
▾ 🔗 Mates
  ◎ Concentric1 (Bracket<1>,Bush
  ⟋ Coincident1 (Bracket<1>,Bush
  ◎ Concentric2 (Bushing<1>,Axle
  ↦ Distance1 (Axle<1>,Bushing<
  ◎ Concentric3 (Axle<1>,Wheel<
  ↦ Distance2 (Wheel<1>,Bushing
  ◎ Concentric4 (Axle<1>,Collar<
  ⟋ Coincident2 (Wheel<1>,Collar

Activity: Insert and Mate the Sixth Component - 2MM Set Screw.

Insert the 2MM Set Screw into the Collar.

110) Click **Insert Component** from the Assembly toolbar. The Insert Component PropertyManager and the Window Open dialog box is displayed.

111) Double-click the **2MM Set Screw** part from SOLIDWORKS 2024\FLY WHEEL folder.

112) Click a **position** as illustrated. The 2MM Set Screw is displayed.

Insert a Concentric Mate between the inside cylindrical face of the Collar and the outside face of the 2MM Set Screw. If needed, apply the Anti-Aligned option.

113) **Zoom-in** on the Collar and the 2MM Set Screw part.

114) Click the **Mate** tool from the Assembly toolbar. The Mate PropertyManager is displayed.

Pin the Mate PropertyManager.

115) Click the **Keep Visible** icon from the Mate PropertyManager.

116) Click the **Standard Mate** tab.

117) Click the **inside cylindrical face** of the Collar. Face<1>@Collar-1 is displayed in the Mate Selections box.

118) Click the **outside cylindrical face** of the **2MM Set Screw** part. Face<2>@2MM Set Screw is displayed in the Mate Selections box.

119) If needed, click **Flip Mate Alignment (Anti-Aligned)**. Concentric mate is selected by default.

120) Click **OK** from the Concentric Mate PropertyManager.

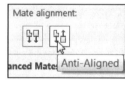

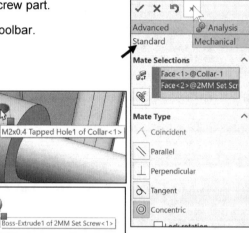

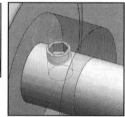

Insert a Tangent Mate between the flat face of the 2MM Set Screw and the cylindrical face of the Axle.

121) **Zoom-in** on the flat face of the 2MM Set Screw part as illustrated.

122) Drag the **2MM Set Screw** to view the bottom flat face.

123) Click the **flat face** of the 2MM Set Screw as illustrated.

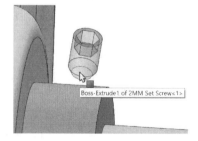

124) Click the **cylindrical face** of the
Axle. Tangent ⟨⟩ mate is
selected by default.

125) Click **OK** ✔ from the Tangent1
Mate PropertyManager.

126) Un-Pin the Mate
PropertyManager. Click the
Keep Visible icon as illustrated.

127) Click **OK** ✔ from the Mate
PropertyManager. View the
created mates in the Mates
folder.

Fit the Assembly to the Graphics window.
128) Press the **f** key.

Display a Trimetric view.

129) Click **Trimetric view** 🔷 from the Heads-up View
toolbar.

Save the model.
130) Click **Save** 💾.

131) Click **Rebuild** and **save**. View the results.

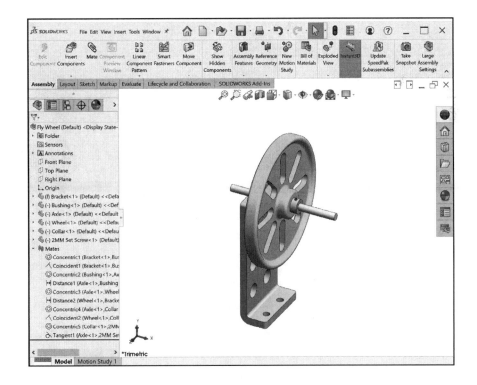

Rotate the Axle. The Collar moves separately about the Axle. The Wheel moves separately about the Axle.

Align the Wheel and Collar so they rotate with respect to the Axle. Insert a Coincident mate between the three planes (Top Plane) for proper assembly movement.

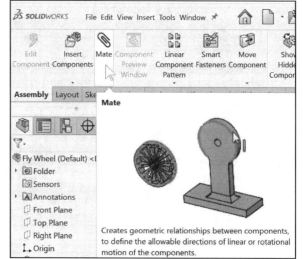

Insert a Coincident mate using the multiple mate mode between the Top Plane of the Axle, Wheel, and Collar.

132) Click the **Mate** 🖉 tool from the Assembly toolbar. The Mate PropertyManager is displayed. The Standard mate tab is selected by default.

133) Expand the fly-out FeatureManager from the Graphics window.

134) Expand the Axle component from the fly-out FeatureManager.

135) Click **Top Plane** from the fly-out FeatureManager.

136) Click **Multiple mate mode** 🖉 from the Mate Selections box. The Components Reference box is displayed.

137) Expand the Wheel component from the fly-out FeatureManager.

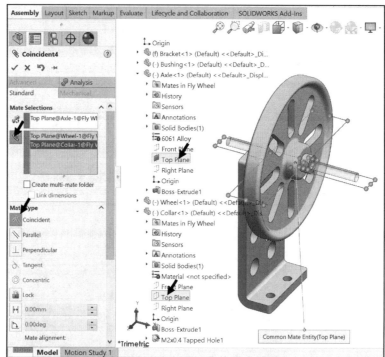

138) Click **Top Plane** from the fly-out FeatureManager.

139) Expand the Collar component from the fly-out FeatureManager.

140) Click **Top Plane** from the fly-out FeatureManager. Coincident 🖊 mate is selected by default.

141) Click **OK** ✔ from the Coincident
Mate PropertyManager.

142) **Rotate** the Wheel and view the
results.

143) **Rotate** the Wheel back to the
original orientation where the 2MM
Set Screw part is on top.

Display a Trimetric view.

144) Click **Trimetric view** 📦 from the
Heads-up View toolbar.

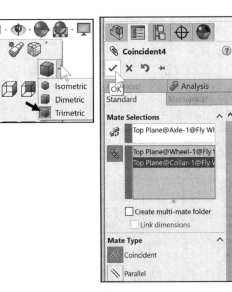

Save the model.

145) Click **Save** 💾.

146) Click **Save** and **Rebuild**.

Exploded View

An Exploded view shows an assembly's
components spread out but positioned to
show how they fit together when assembled.

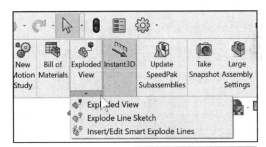

Create exploded views by selecting and
dragging parts in the Graphics area and
creating one or more explode steps. **Note**:
You can also input a dimension between the
components or sub-assemblies.

In exploded views you can:

- Evenly space exploded stacks of components (hardware,
  washers and so on).

- Attach a new component to the existing explode steps of
  another component. This is useful if you add a new part to an
  assembly that already has an exploded view.

- If a subassembly has an exploded view, reuse that view in a
  higher-level assembly.

- Add explode lines to indicate component relationships.

In the next section, create an Exploded view of the Fly Wheel assembly.

Activity: Create an Exploded View of the Fly Wheel Assembly.

Create an Exploded View.

147) Click the **Assembly** tab from the CommandManager.

148) Click the **Exploded View** icon. The Explode PropertyManager is displayed. View your options.

149) Click the **Regular step (translate and rotate)** button.

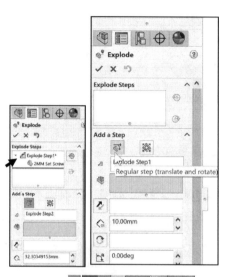

Create Explode Step1 with the 2MM Set Screw.

150) **Zoom-in** on the 2MM Set Screw in the Graphics window.

151) Click the **2MM Set Screw**. A Manipulator icon is displayed.

152) **Zoom-out** to view the assembly.

153) Drag the **Manipulator** icon upward - approximately 30mm using the Graphics window ruler.

154) Release the **Manipulator** icon.

155) Click **Done**. Explode Step1 is created.

Create Explode Step2. Select the Collar and 2MM Set Screw.

156) Click the **Collar**. A Manipulator icon is displayed.

157) Hold the **Ctrl** key down.

158) Click the **2MM Set Screw**.

159) Release the **Ctrl** key.

160) **Zoom-out** to view the Fly Wheel assembly.

161) Drag the **Manipulator** icon outward as illustrated - approximately 140mm using the Graphics window ruler. **Click reverse direction** if needed. **Note:** You can also input 140mm directly into the Explode Distance box, and then click Add Step.

162) Release the **Manipulator** icon.

163) Click **Done**. Create Explode Step2 is created.

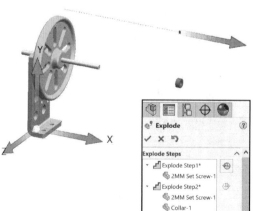

Explode Step3.

164) Click the **Wheel** from the Graphics window. A Manipulator icon is displayed.

165) Drag the **Manipulator** icon to the right of the assembly. Enter 120mm directly into the Explode Distance box, and then click Add Step.

166) Release the **Manipulator** icon.

167) Click **Done**. Explode Step3 is created.

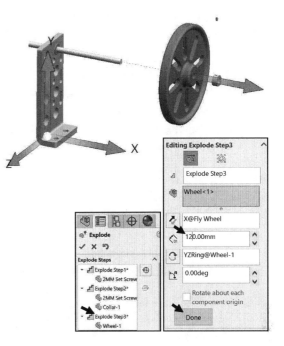

Explode Step4.

168) Click the **Axle** from the Graphics window. A Manipulator icon is displayed.

169) Drag the **Manipulator** icon to the right of the assembly - approximately 40mm using the Graphics window ruler.

170) Release the **Manipulator** icon.

171) Click **Done**. Explode Step4 is created.

Create Explode Step5.

172) Click the **Bushing** from the Graphics window. A Manipulator icon is displayed.

173) Drag the **Manipulator** icon to the left of the assembly - approximately 40mm - using the Graphics window ruler.

174) Click **Done**. Explode Step5 is created.

175) Click **OK** ✔ from the Explode PropertyManager.

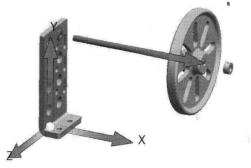

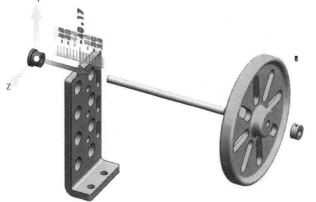

View the Animate collapse of the Exploded Fly Wheel Assembly.

176) **Expand** the Default folder in the ConfigurationManager as illustrated.

177) Right-click **Exploded View1**.

178) Click **Animate collapse**. The Animation Controller is displayed.

179) Click **Playback Mode: Loop**. View the results in the Graphics window.

180) Click **Pause** .

181) Click **End** .

182) **Close** the Animation Controller.

Return to the Assembly FeatureManager.

183) Click the **FeatureManager Design Tree** icon.

Save the model. Make sure you are saving the Assembly document and all reference components in the save file folder.

184) Click **Save** .

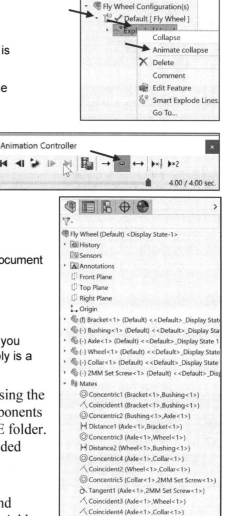

Close all documents.

185) Click **Window**, **Close All** from the Main menu. Are you finished with the assembly? The Fly Wheel assembly is a sub-assembly to the final Stirling Engine assembly.

In the next section, create the final Stirling Engine using the Fly Wheel assembly along with other provided components from the SOLIDWORKS 2024\STIRLING ENGINE folder. Use the components and sub-assemblies in the provided folder.

CommandManager and FeatureManager tabs and folder files will vary depending on system setup and Add-ins.

Activity: Create the Final Stirling Engine Assembly.

Open the Stirling Engine Assembly.

186) Click **Open** from the Main menu.

187) Browse to the **SOLIDWORKS 2024\STIRLING ENGINE** folder.

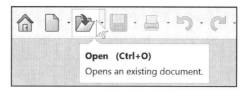

188) Double-click the **Stirling Engine** assembly. The Stirling Engine is displayed in the Graphics window.

189) **Review** the Feature Manager. The Feature Manager is made up of components. Components are sub-assemblies and parts. There are two sub-assemblies: Main Cylinder and Power. Both are displayed in a flexible state.

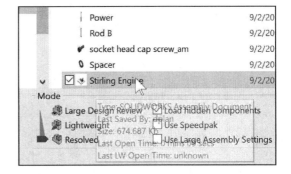

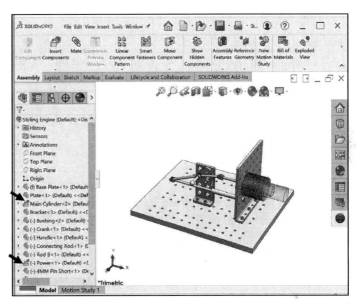

The Plate and Bracket are designed for flexibility to support other components in the future.

Activity: Hide the Plate Component.

Hide the Plate component.

190) Right-Click **Plate** from the Assembly FeatureManager.

191) Click **Hide Components** from the Context pop-up toolbar.

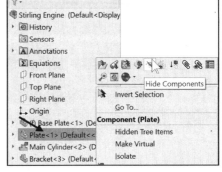

Activity: Insert and Rotate the Fly Wheel Assembly.

Insert the Fly Wheel assembly.

192) Click **Insert Component** 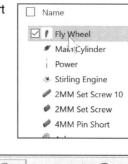 from the Assembly toolbar. The Insert
Component PropertyManager and the Window Open
dialog box is displayed.

193) Double-click the **Fly Wheel** assembly from the
SOLIDWORKS 2024\STIRLING ENGINE folder.

194) Click a **position** above the assembly. The Fly Wheel is
free to translate. Drag the Fly Wheel behind the Stirling
Engine assembly.

Rotate the Fly Wheel assembly.

195) Click **Rotate Component** from the Move Component
drop-down menu. The Rotate Component
PropertyManager is displayed.

196) **Rotate** the Fly Wheel assembly as illustrated.

197) Click **OK** ✔ from the Rotate Component
PropertyManager.

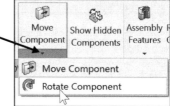

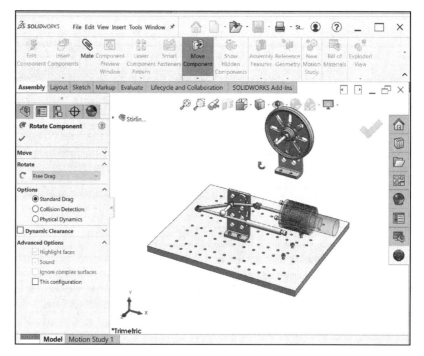

Activity: Mate the Fly Wheel Assembly.

Insert a Concentric mate between the cylindrical face of the Axle and the cylindrical face of the Crank hole.

198) Click the **cylindrical face** of the Axle.

199) Click **Front view** .

200) **Rotate** and **Zoom in** on the Crank hole.

201) Hold the **Ctrl** key down.

202) Click the **inside cylindrical face** of the Crank hole as illustrated.

203) Release the **Ctrl** key. The Mate Pop-up menu is displayed.

204) Click **Concentric** from The Mate Pop-up menu. A Concentric mate locates the selected items so they can share the same center point.

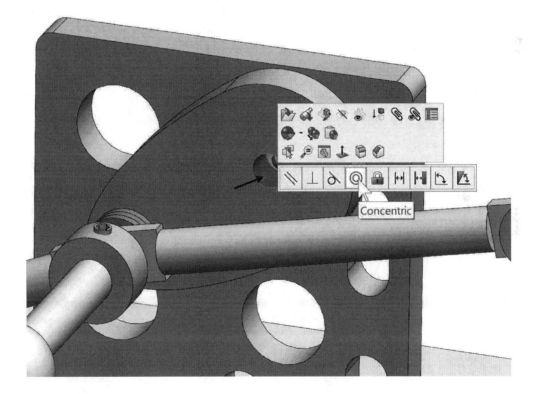

Insert a Concentric mate between the cylindrical face of the Bracket hole and the cylindrical face of the Base Plate hole directly aligned above with the Axle.

205) Rotate and **Zoom** in as illustrated. If needed move the Fly Wheel assembly.

206) Click the inside **cylindrical face** of Bracket hole as illustrated.

207) Hold the **Ctrl** key down.

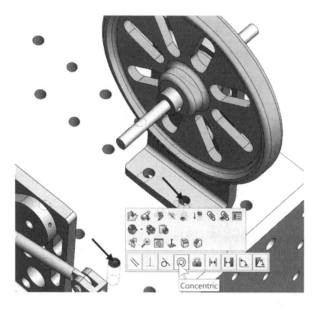

208) Click the **cylindrical face** of the Base Plate hole as illustrated.

209) Release the **Ctrl** key. The Mate Pop-up menu is displayed.

210) Click **Concentric** ◎ from The Mate Pop-up menu.

The Fly Wheel Bracket and the front Bracket are aligned. No additional spacers are needed to locate the Fly Wheel Bracket to the Base Plate.

View the length of the Axle.
211) Double-click on the **Axle** in the Graphics window. The Axle length is 80mm. If needed, rotate the model to view the dimension. The Axle is too short and does not go through the Crank. The Axle ends at the back face of the Bushing.

212) Click **inside** the Graphics window.

In the next section, apply the Measure tool to locate the needed length of the Axle.

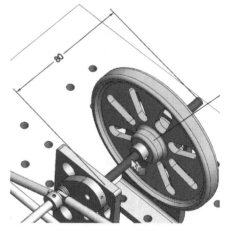

Apply the Measure tool.

213) Press the **f** key to fit the model to the Graphics window.

214) Click the **Evaluate** tab from the CommandManager.

215) Click the **Measure** 🔍 tool. The Measure dialog box is displayed.

216) **Zoom in** and **Rotate** on the Bushing and Crank components as illustrated.

217) Click the **back face** of the Bushing.

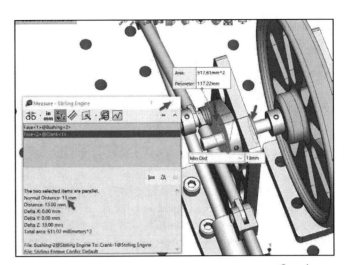

218) Click the **front face** of the Crank. The Normal Distance between the two faces is 13mm.

219) **Close** ❌ the Measure dialog box.

Fit the model to the Graphics window.

220) Press the **f** key.

Modify the Axle length.

221) Double click on the **Axle** in the Graphics window. View the dimensions.

222) Open the **Axle** part in the Assembly.

223) Double-click on **Boss-Extrude1**.

224) Modify the overall length from **80** to **93**.

225) **Save** the Axle.

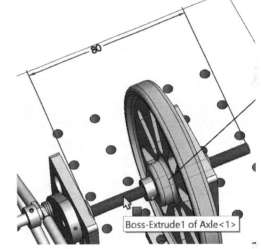

Boss-Extrude1 of Axle<1>

226) **Return** to the assembly. The Axle is the correct length.

Display a Trimetric view.

227) Click **Trimetric view** 🔷 from the Heads-up View toolbar.

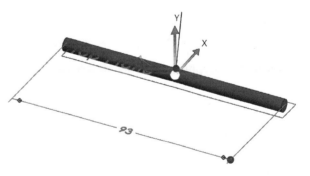

Save the model.

228) Click **Save** 💾.

229) Click **Save All**.

Click and drag the Handle to rotate the Crank. The Crank rotates. The linkage controls the linear motion of the Power Piston.

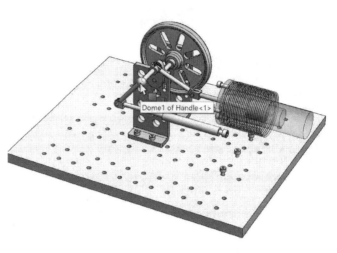

The Main Cylinder Piston translates. The Fly Wheel is rigid and does not translate or rotate. The Fly Wheel is a sub-assembly and used mates to rotate the Axle, Shaft Collar and Wheel. When a component is inserted into an assembly, any mates are solved as rigid.

In the next section, make the Fly Wheel flexible within the main assembly.

Make the Fly Wheel sub-assembly Flexible.

230) Right-click **Fly Wheel** from the Assembly FeatureManager.

231) Click **Make Subassembly Flexible** from the Pop-up Content toolbar.

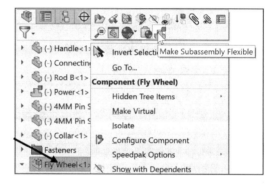

232) Rotate the Wheel in the Graphics window. The Axle of the Fly Wheel assembly is free to rotate. The Crank is free to rotate.

In the next section, apply mates to control the rotation of the Crank and Axle.

Insert a Coincident Mate between the Top Plane of the Axle and the Top Plane of the Crank.

233) Expand the Fly Wheel component from the FeatureManager.

234) Expand the Axle component from the FeatureManager.

235) Click **Top Plane**.

236) Hold the **Ctrl** key down.

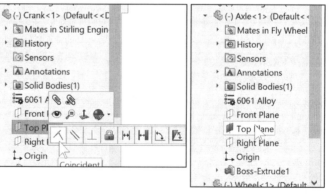

237) Expand the Crank component from the fly-out FeatureManager.

238) Click **Top Plane** from the fly-out FeatureManager.

239) Release the **Ctrl** key. The Mate Pop-up menu is displayed.

240) Click **Coincident** ⚲ mate from the Mate pop-up menu. Note if these two planes were not coincident, a parallel plane could be used to control rotation.

241) **Rotate** the Crank. The Fly Wheel, Crank and Axle all rotate together.

Display the Plate.

242) Right-click **Plate** from the Assembly FeatureManager.

243) Click **Show Components** for the Content toolbar. View the final assembly in the Graphics window.

Display a Trimetric view.

244) Click **Trimetric view** 🟦 from the Heads-up View toolbar.

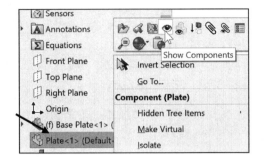

Save the model.

245) Click **Save** 💾.

246) Click **Rebuilt and Save**.

Pack and Go

The Pack and Go tool gathers all related files for a model design (parts, assemblies, drawings, references, design tables, Design Binder content, decals, appearances, scenes and SOLIDWORKS Simulation results) into a single folder or zip file.

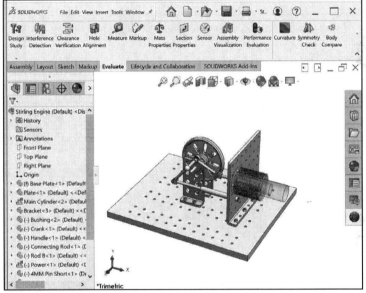

It's one of the best tools to utilize when you are trying to save a large assembly or drawing with references and SOLIDWORKS Toolbox components. **Note**: A new option is to save to a previous older version (2023, 2022).

In the next section apply the Pack and Go tool to save the assembly and all toolbox components. The Pack and Go dialog box list related files to be saved into a folder or zip file.

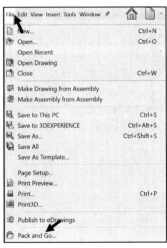

Activity: Pack and Go the Stirling Engine Assembly.

Save the Assembly using Pack and Go to a Zip file.

247) Click **File**, **Pack and Go** from the Main menu bar. The Pack and Go dialog box is displayed. The dialog box lists related files to be saved into a folder or zip file. For additional information see SOLIDWORKS Help.

248) **View** your options.

249) Check **Save to Zip file** as illustrated.

250) Check the **Include Toolbox components**.

251) Check the **Include suppressed components**.

252) **Browse** to the SOLIDWORKS 2024 folder.

253) Enter the **File name**.

254) Click **Save** from the Save As dialog box.

255) Click **Save** from the Pack and Go dialog box. A Zip file is created of the Zip assembly and all reference components.

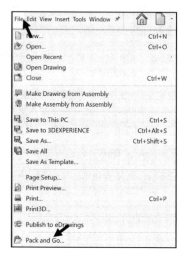

Close all models.

256) Click **Window, Close All** from the Main menu. You are finished with this chapter.

Note: <u>Save to This PC</u>: Saves the document to your local hard drive in the folder which was last opened from. <u>Save to 3DEXPERIENCE</u>: Opens the Save as New dialog box to save the file to a Collaborative space on the **3D**EXPERIENCE platform. The Save to 3DEXPERIENCE option is displayed if you purchased 3DEXPERIENCE education roles and are logged into the **3D**EXPERIENCE platform. <u>Save As</u>: Saves the document to a new file name that becomes the active document without saving the original document. <u>Save All</u>: Saves all the open files that have been modified since they were last saved.

Summary

You established a SOLIDWORKS session and created two new assemblies with user defined document properties:

- Fly Wheel

- Stirling Engine

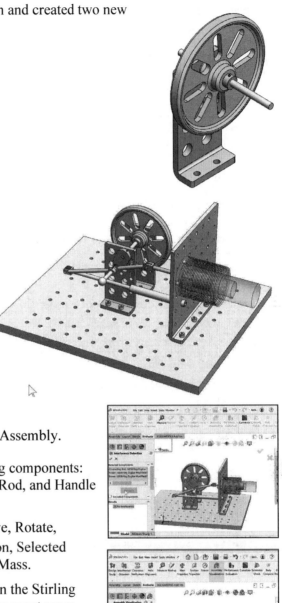

You inserted the following Standard mate types: Coincident, Concentric, Distance and Tangent.

You utilized the following assembly tools: Insert Component, Suppress, Un-suppress, Mate, Move Component, Rotate Component, Hide, Show, Flexible, Ridge and Multiple mate mode.

You created an Exploded View with animation and applied the Measure and Mass Properties tool.

You saved the final assembly using Pack and Go.

In Chapter 4, address clearance, interference, static and dynamic behavior of the Stirling Engine Modified Assembly.

Verify the behavior between the following components: Power Piston, Power Clevis, Connecting Rod, and Handle in the assembly.

Apply the following assembly tools: Move, Rotate, Collision Detection, Interference Detection, Selected Components, Edit Feature and Center of Mass.

Utilize the Assembly Visualization tool on the Stirling Engine Modify assembly and sort by component mass.

Create a new Coordinate System on the Stirling Engine Modify assembly relative to the default origin.

Create, run, and save a Motion Study.

Questions

1. Identify the three default reference planes in SOLIDWORKS.

2. True or False. Sketches are created only on the Front plane.

3. True or False. An assembly is a document that contains two or more parts. An assembly inserted into another assembly is called a sub-assembly.

4. True or False. Determine the static and dynamic behavior of mates in each sub-assembly before creating the top-level assembly.

5. Describe a Distance mate.

6. Describe a Coincident mate.

7. Identify three Standard mate types in SOLIDWORKS.

8. True or False. A fixed component cannot move in a SOLIDWORKS assembly.

9. Describe the procedure to remove the fix state (f) of a component in an assembly.

10. Describe the procedure to rotate a component in an assembly.

11. Describe the procedure to resolve a lightweight component in an assembly.

12. In an assembly, each component has_____# degrees of freedom.

13. True or False. A Part Template is the foundation for an SOLIDWORKS assembly document.

14. True or False. If you delete a mate and then recreate it, the mate number will be different (increase).

Exercises

Exercise 3.1: HEX-NUT Part

Create an ANSI, IPS HEX-NUT part. Apply 6061 Alloy material. Apply the following dimensions:

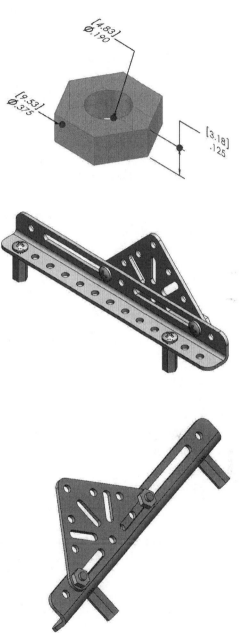

- Depth: .125 in, [3.18].

- Inside hole diameter: .190in, [4.83].

- Outside diameter: .375in, [9.53].

Use the Top Plane as the Sketch plane.

Exercise 3.2: FRONT-SUPPORT-2 Assembly

Create an ANSI, IPS FRONT-SUPPORT-2 assembly.

- Name the new assembly FRONT-SUPPORT-2.

- Insert the FRONT-SUPPORT assembly. The FRONT-SUPPORT assembly is provided in the **Chapter 3 Homework folder**.

- Fix the FRONT-SUPPORT assembly to the Origin.

- Insert the first HEX-NUT (Exercise 3.1) into the FRONT-SUPPORT-2 assembly.

- Insert a Concentric mate and Coincident mate.

- Insert the second HEX-NUT part.

- Insert a Concentric mate and Coincident mate.

You can also insert a Parallel mate between the HEX-NUT parts and the FRONT-SUPPORT assembly.

Exercise 3.3: Weight-Hook Assembly

Create an ANSI, IPS Weight-Hook assembly. The Weight-Hook assembly has two components: WEIGHT and HOOK.

- Create a new ANSI, IPS assembly document.

- Insert the WEIGHT part from the Chapter 3 Homework folder.

- Fix the WEIGHT to the Origin in the FeatureManager.

- Insert the HOOK part from the Chapter 3 Homework folder into the assembly.

- Insert a Concentric mate between the inside top cylindrical face of the WEIGHT and the cylindrical face of the thread. Concentric is the default mate.

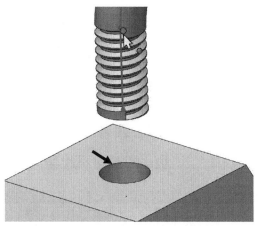

- Insert the first Coincident mate between the top edge of the circular hole of the WEIGHT and the **top point** of the thread as illustrated. The HOOK can rotate in the WEIGHT.

- Fix the position of the HOOK. Insert the second Coincident mate between the Right Plane of the WEIGHT and the Right Plane of the HOOK. Coincident is the default mate.

- Expand the Mates folder and view the created mates.

- Calculate the Mass and Volume of the assembly.

- Identify the Center of Mass for the assembly.

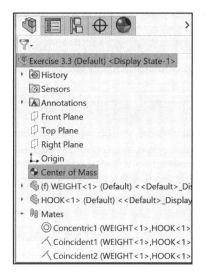

Exercise 3.4: Weight-Link Assembly

Create an ANSI, IPS Weight-Link assembly. The Weight-Link assembly has two components and a sub-assembly: Axle component, FLATBAR component and the Weight-Hook sub-assembly that you created in a previous exercise.

- Create a new assembly document. Insert the Axle part from the Chapter 3 Homework folder.

- Fix the Axle component to the Origin.

- Insert the FLATBAR part from the Chapter 3 Homework folder. Rotate as needed.

- Insert a Concentric mate between the Axle cylindrical face and the FLATBAR inside face of the top circle.

- Insert a Coincident mate between the Front Plane of the Axle and the Front Plane of the FLATBAR.

- Insert a Coincident mate between the Right Plane of the Axle and the Top Plane of the FLATBAR.

- Position the FLATBAR as illustrated.

- Insert the Weight-Hook sub-assembly that you created in Exercise 3.3.

- Insert a Tangent mate between the inside bottom cylindrical face of the FLATBAR and the top circular face of the HOOK, in the Weight-Hook assembly. Tangent mate is selected by default. Click Flip Mate Alignment if needed.

- Insert a Coincident mate between the Front Plane of the FLATBAR and the Front Plane of the Weight-Hook sub-assembly. Coincident mate is selected by default. The Weight-Hook sub-assembly is free to move in the bottom circular hole of the FLATBAR.

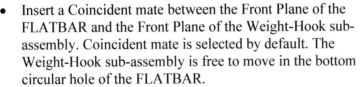

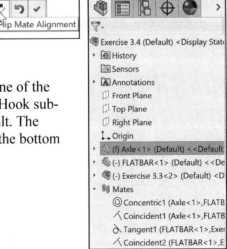

- **Exercise 3.5: Binder Clip**

- Create a simple Gem® binder clip.

- Create an ANSI - IPS assembly.
  Create two components - Binder and
  Binder Clip.

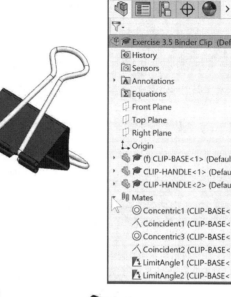

- Apply material (plan carbon steel) to
  each component and address all
  needed mates. Think about where
  you would start.

Note: The cuts in the main body are made before
folding the metal and strengthening the material
during manufacturing. You do not need to model
this exactly. The key to this exercise is to apply
Advance distance mates (Limit mates) to limit the
movement of the handle.

- What is the Base Sketch for each component?

- What are the dimensions? Measure all
  dimensions (approximately) from a small or
  large Gem binder clip.

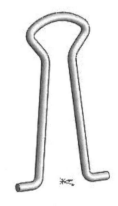

- View SOLIDWORKS Help or Chapter 5 on the
  Swept Base feature to create the Binder Clip
  handle.

Determine the static and dynamic behavior of
mates in each sub-assembly before creating the top
level assembly.

**Exercise 3.6: Limit Mate
(Advanced Mate)**

- Copy all files from the
  Chapter 3
  Homework\Limit Mate
  folder to your hard
  drive.

- Open the Limit Mate
  assembly. View the created mates.

- Create the final assembly with all needed mates for proper
  movement.

- Insert a Distance (Advanced Limit Mate) to restrict the
  movement of the Slide Component - lower and upper
  movement.

A Distance (Limit Mate) is an
Advanced Mate type. Limit mates
allow components to move within a
range of values for distance and
angle. You specify a starting distance
or angle as well as a maximum and
minimum value.

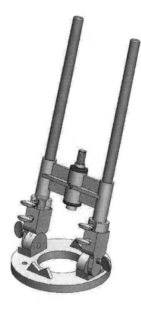

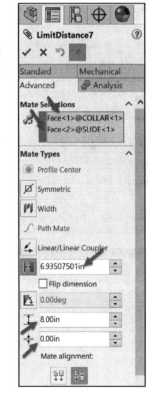

- Save the model and move the
  slide to view the results in the
  Graphics window.

- Think about how you would use
  this mate type in other
  assemblies.

Exercise 3.7: Screw Mate (Mechanical Mate)

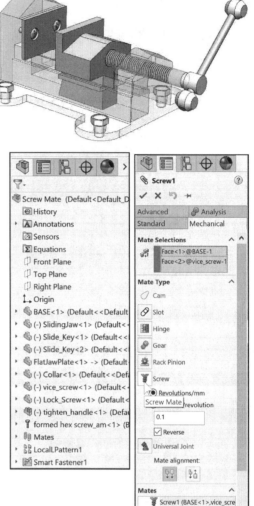

- Copy the Chapter 3 Homework\Screw Mate folder to your hard drive.

- Open the Screw Mate assembly.

- Insert a Screw mate between the inside Face of the Base and the Face of the vice and any other mates that are required. A Screw is a Mechanical Mate type.

- Calculate the total mass and volume of the assembly.

A Screw mate constrains two components to be concentric and also adds a pitch relationship between the rotation of one component and the translation of the other. Translation of one component along the axis causes rotation of the other component according to the pitch relationship. Likewise, rotation of one component causes translation of the other component. Use SOLIDWORKS Help if needed.

- Rotate the handle and view the results. Think about how you would use this mate type in other assemblies.

Use the Select Other tool (See SOLIDWORKS Help) to select faces and edges that are hidden in an assembly.

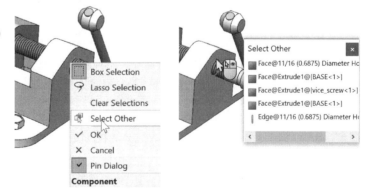

Exercise 3.8: Angle Mate

- Copy the Chapter 3 Homework\Angle Mate folder to your hard drive.

- Open the Angle Mate assembly.

- Move the Handle in the assembly. The Handle is free to rotate.

- Set the angle of the Handle.

- Insert an Angle mate (165 degree) between the Handle and

  the Side of the valve using Planes. An Angle mate places the selected items at the specified angle to each other. The Handle has a 165 degree Angle mate to restrict flow through the valve.

Think about how you would use this mate type in other assemblies.

Exercise 3.8A: Angle Mate (Cont:)

- Create two end caps (lids) for the ball valve using the Top-down Assembly method. Note: The Reference-In-Content symbols in the FeatureManager.

- Modify the Appearance of the body to observe the change and enhance visualization.

- Apply the Select-other tool to obtain access to hidden faces and edges.

Exercise 3.9: Symmetric Mate (Advanced Mate)

- Copy the Chapter 3 Homework\Symmetric Mate folder to your hard drive.

- Open the Symmetric Mate assembly.

- Insert a Symmetric Mate between the two Guide Rollers.

- Apply LimitMates (Advanced Distance mate) to limit movement of the two Guide Rollers.

- Think about how you would use this mate type in other assemblies?

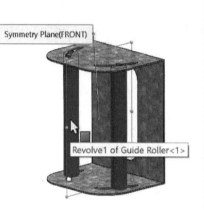

Exercise 3.10: Gear Mate (Mechanical mate)

View the ppt presentation and motion file on Gears located in the Chapter 3 Homework\Gears folder.

Create the Gear Mate assembly per the ppt presentation. Create the **Base Plate** and **two Shafts** per the ppt presentation.

Use the SOLIDWORKS toolbox to insert the needed gear components. You create the **Base Plate**, and **Shafts**.

Insert all needed mates for proper movement.

Gear mate: Forces two components to rotate relative to one another around selected axes. The Gear mate provides the ability to establish gear type relations between components without making the components physically mesh.

SOLIDWORKS provides the ability to modify the gear ratio without changing the size of the gears. *Align the components* before adding the Mechanical gear mate.

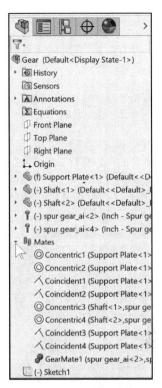

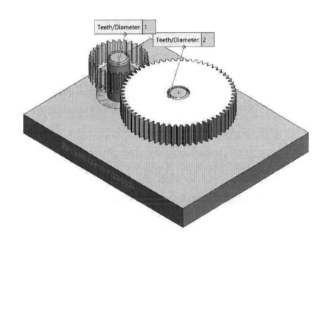

Exercise 3.10A: Slot Mate (Center in Slot option)

Copy the Chapter 3 Homework\Slot Mate folder to your hard drive.

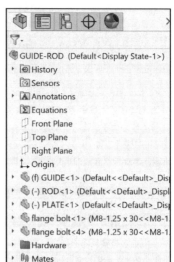

Open the GUIDE-ROD assembly. Create two Slot mates for the two flange bolts. You can mate bolts to straight or arced slots and you can mate slots to slots. Use the Quick Mate procedure.

Insert a Slot mate (Center in Slot option) between the first flange bolt and the inside face of the right slot.

Insert the second slot mate (Center in Slot option) between the second flange bolt and the inside face of the left slot.

Insert a Coincident mate between the top face of the GUIDE and the bottom face of the second flange bolt cap.

Insert a Coincident mate between the top face of the GUIDE and the bottom face of the first flange bolt cap.

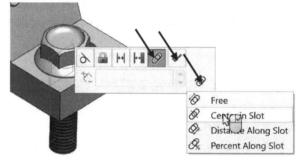

Insert a parallel mate between the front face of the first hex head flange bolt and the front of the GUIDE.

Inset a parallel mate between the front face of the second hex head flange bolt and the front face of the GUIDE.

Expand the Mates folder. View the created mates.

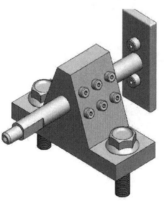

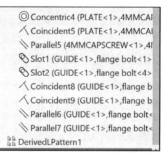

Exercise 3.11: Slider Part

Create the part from the illustrated A-ANSI - MMGS Third Angle Projection drawing below: Front, Top, Right and Isometric view.

Note: The location of the Origin (shown in an Isometric view).

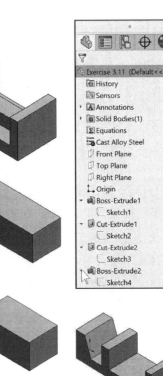

- Apply Cast Alloy steel for material.

- The part is symmetric about the Front Plane.

- Apply Mid Plane for End Condition in Boss-Extrude1.

- Apply Through All for End Condition in Cut-Extrude1.

- Apply Through All for End Condition in Cut-Extrude2.

- Apply Up to Surface for End Condition in Boss Extrude2.

- Calculate the Volume of the part and locate the Center of mass.

Think about the steps that you would take to build the model. Do you need the Right view for manufacturing? Does it add any important information?

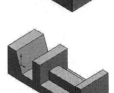

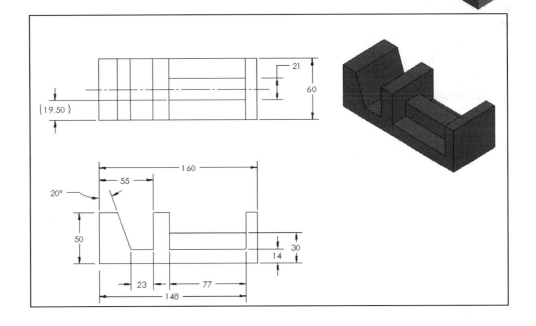

Exercise 3.12: Cosmetic Thread Part

- **Apply a Cosmetic thread:** ANSI 1/4-20-1 UNC 2A. A cosmetic thread represents the inner diameter of a thread on a boss or the outer diameter of a thread.

- Copy the Cosmetic thread part from the Chapter 3 Homework folder to your hard drive.

- Open the part.

- Create a Cosmetic thread. Produce the geometry of the thread.

- Click the bottom edge of the part as illustrated.

- Click Insert, Annotations, Cosmetic Thread from the Menu bar menu. View the Cosmetic Thread PropertyManager. Edge<1> is displayed.

- Select ANSI Inch for Standard.

- Select Machine Threads for Type.

- Select ¼-20 for Size.

- Select Blind for End Condition.

- Enter 1 for depth.

- Select 2A for Thread class.

- Click OK 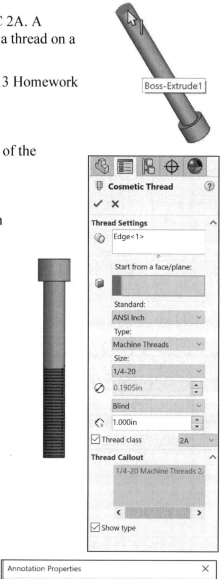 from the Cosmetic Thread FeatureManager.

- Expand the FeatureManager. View the Cosmetic Thread feature. If needed, right-click the Annotations folder, click Details.

- Check the Cosmetic threads and Shaded cosmetic threads box.

- Click OK. View the cosmetic thread on the model.

Exercise 3.13: Hole - Block

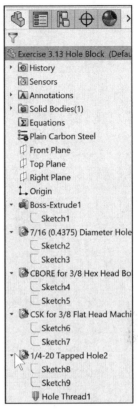

- Create the Hole-Block ANSI - IPS part using the Hole Wizard feature as illustrated. Create the Hole-Block part on the Front Plane.

- The Hole Wizard 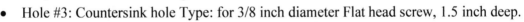 feature creates either a 2D or 3D sketch for the placement of the hole in the FeatureManager.

- You can consecutively place multiple holes of the same type. The Hole Wizard creates 2D sketches for holes unless you select a non-planar face or click the 3D Sketch button in the Hole Position PropertyManager.

Hole Wizard creates two sketches. A sketch of the revolved cut profile of the selected hole type and the other sketch of the center placement for the profile. Both sketches should be fully defined.

- Create a rectangular prism 2 inches wide by 5 inches long by 2 inches high. On the top surface of the prism, place four holes, 1 inch apart.

- Hole #1: Simple Hole Type: Fractional Drill Size, 7/16 diameter, End Condition: Blind, 0.75 inch deep.

- Hole #2: Counterbore hole Type: for 3/8 inch diameter Hex bolt, End Condition: Through All.

- Hole #3: Countersink hole Type: for 3/8 inch diameter Flat head screw, 1.5 inch deep.

- Hole #4: Tapped hole Type, Size ¼-20, End Condition: Blind -1.0 inch deep.

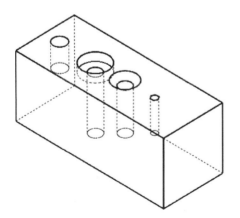

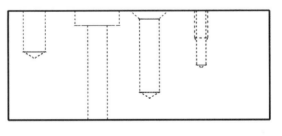

Exercise 3.14: Hole Wizard Part

- Apply the 3D sketch placement method as illustrated in the FeatureManager. Insert and dimension a hole on a cylindrical face.

- Copy the Hole Wizard 3-14 part from the Chapter 3 Homework folder to your hard drive.

- Open the Hole Wizard 3-14 part.

Note: With a 3D sketch, press the Tab key to move between planes.

- Click the Hole Wizard 📷 Features tool. The Hole Specification PropertyManager is displayed.

- Select the Counterbore Hole Type.

- Select ANSI Inch for Standard.

- Select Socket Head Cap Screw for fastener Type.

- Select 1/4 for Size. Select Normal for Fit. Select Through All for End Condition.

- Enter .100 for Head clearance in the Options box. Click the Positions Tab. The Hole Position PropertyManager is displayed.

- Click the 3D Sketch button. SOLIDWORKS displays a 3D interface with the Point ✏️ XY tool active.

💡 When the Point tool is active, wherever you click, you will create a point.

- Click the cylindrical face of the model as illustrated. The selected face is displayed in blue.

- Insert a dimension between the top face and the Sketch point.

- Click the Smart Dimension ↖ Sketch tool.

- Click the top flat face of the model and the sketch point.

- Enter .25in.

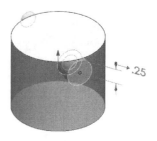

- Locate the point angularly around the cylinder. Apply construction geometry.

- Activate the Temporary Axes. Click View, Hide/show, check the Temporary Axes box from the Menu bar toolbar.

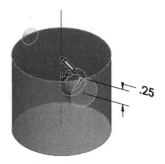

- Click the Line ✎ Sketch tool. Note: 3D sketch is still activated.

- Ctrl+click the top flat face of the model. This moves the red space handle origin to the selected face. This also constrains any new sketch entities to the top flat face. Note the mouse pointer ⬚ ⤢ icon.

- Move the mouse pointer near the center of the activated top flat face as illustrated. View the small black circle. The circle indicates that the end point of the line will pick up a Coincident relation.

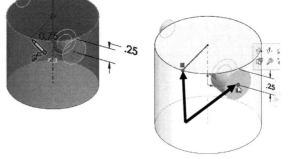

- Click the center point of the circle.

- Sketch a line so it picks up the **AlongZ sketch relation**. The cursor displays the relation to be applied. **This is a very important step. If needed, create a AlongZ sketch relation**.

- Create an AlongY sketch relation between the centerpoint of the hole on the cylindrical face and the endpoint of the sketched line as illustrated. The sketch is fully defined.

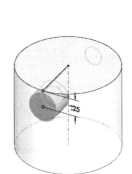

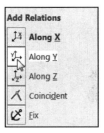

Add Relations	
↗	**Along X**
↘	Along Y
↗	Along Z
⤢	Coincident
✗	Fix

- Click OK ✔ from the Properties PropertyManager. Click OK ✔ from the Hole Position PropertyManager.

- Expand the FeatureManager and view the results. The two sketches are fully defined. One sketch is the hole profile, the other sketch is to define the position of the feature. Close the model.

Exercise 3.15: Counter Weight Assembly

Create the Counter Weight assembly as illustrated using SmartMates and Standard mates.

Copy the Chapter 3 Homework\Counter-Weight folder to your local hard drive.

Create a new assembly. The Counter Weight assembly consists of the following items:

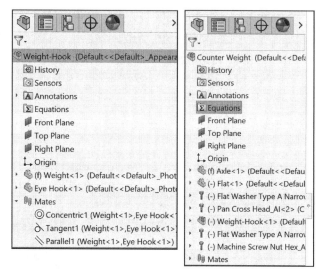

- Weight-Hook sub-assembly.

- Weight.

- Eye Hook.

- Axle component (f). Fixed to the origin.

- Flat component.

- Flat Washer Type A (from the SOLIDWORKS Toolbox).

- Pan Cross Head Screw (from the SOLIDWORKS Toolbox).

- Flat Washer Type A (from the SOLIDWORKS toolbox).

- Machine Screw Nut Hex (from the SOLIDWORKS Toolbox).

Apply SmartMates with the Flat Washer Type A Narrow_AI, Machine Screw Nut Hex_AI and the Pan Cross Head_AI components.

Use a Distance mate to fit the Axle in the middle of the Flat. Note a Symmetric mate could replace the Distance mate. Think about the design of the assembly. Apply all needed Lock mates.

The symbol (f) represents a fixed component. A fixed component cannot move and is locked to the assembly Origin.

Exercise 3.16: Gear Mate

Insert a Mechanical Gear mate between two gears. Gear mates force two components to rotate relative to one another about selected axes. Valid selections for the axis of rotation for gear mates include cylindrical and conical faces, axes, and linear edges.

- Copy the Chapter 3 Homework folder to your hard drive.

- Open Mechanical Gear Mate from the Chapter 3 Homework\Gear Mate folder.

- Click and drag the right gear.

- Click and drag the left gear. The gears move independently.

- Click the Mate ✎ tool from the Assembly tab. The Mate PropertyManager is displayed. Click the Mechanical tab.

- Clear all selections.

- Align the components before adding the Mechanical mate. To align the two components, select a Front view and move them manually.

- Select the inside cylindrical face of the left hole.

- Select the inside cylindrical face of the right hole.

- Click Gear ✪ mate from the Mechanical Mates box. The GearMate1 PropertyManager is displayed. Accept the default settings.

- Click OK ✔ from the GearMate1 PropertyManager.

- Expand the Mates folder. View the created GearMate1.

- Rotate the Gears in the Graphics window.

- View the results.

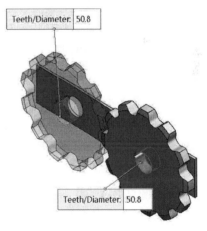

Exercise 3.17: Cam Mate

Create a Tangent Cam follower mate and a Coincident Cam follower mate. A cam-follower mate is a type of Tangent or Coincident mate. It allows you to mate a cylinder, plane or point to a series of tangent extruded faces, such as you would find on a cam.

Create the profile of the cam from lines, arcs and splines as long as they are tangent and form a closed loop.

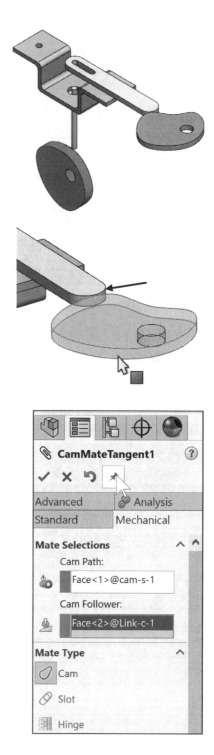

- Copy the Chapter 3 Homework folder to your hard drive.

- Open Cam Mate from the from the Chapter 3 Homework\Cam Mate folder. View the model. The model contains two cams.

- Expand the Mates folder to view the existing mates.

Insert a tangent Cam follower mate.

- Click the Mate ✎ tool from the Assembly tab.

- Click the Mechanical tab.

- Pin the Mate PropertyManager.

- Click Cam Follower ⬭ mate.

- Click inside of the Cam Path box.

- Click the outside face of cam-s<1> as illustrated.

- Click the face of Link-c<1> as illustrated. Face<2>@Link-c-1 is displayed.

- Click OK ✔ from the CamMateTangent1 PropertyManager.

Insert a Coincident Cam follower mate.

- Click Cam Follower ⟨icon⟩ mate from the Mechanical Mates box.

- Click the outside face of cam-s2<1>.

- Click the vertex of riser<1> as illustrated. Vertex<1>@riser-1 is displayed.

- Un-pin the Mate PropertyManager.

- Click OK ✔ from the CamMateCoincident1 PropertyManager.

- Expand the Mates folder. View the new created mates.

- Display an Isometric view.

- Rotate the cams.

- View the results in the Graphics window.

When creating a cam follower mate, make sure that your Spline or Extruded Boss/Base feature, which you use to form the cam contact face, does nothing but form the face.

In the next exercise of the spur gear and rack, the gear's pitch circle must be tangent to the rack's pitch line, which is the construction line in the middle of the tooth cut.

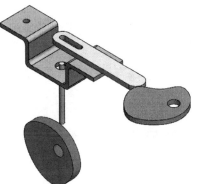

Exercise 3.18: Rack Pinion Mate

The Spur gear and rack components for this exercise were obtained from the SOLIDWORKS Toolbox folder marked Power Transmission. Insert a Mechanical Rack Pinion mate.

- Copy the Chapter 3 Homework folder to your hard drive.

- Open Rack Pinion Mate from the from the Chapter 3 Homework\Rack Pinion Mate folder.

- View the displayed: (TooCutSkeSim and ToothCutSim). If needed, display the illustrated sketches to mate the assembly.

The Diametral Pitch, Pressure Angle and Face Width values should be the same for the spur gear as the rack.

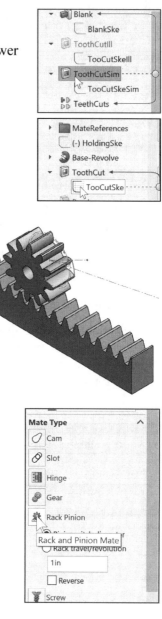

💡 Tangency is required between the spur gear and rack.

- Expand the Mates folder to view the existing mates. Four Coincident mates and a Distance mate were created in the assembly.

Insert a Rack Pinion mate.

- Click the Mate 📎 tool from the Assembly tab.

- Click the Mechanical mate tab.

- Click Rack Pinion 🜨 mate from the Mechanical Mates box.

- Mate the rack. Make sure the teeth of the spur gear and rack are meshing properly; they are not interfering with each other. It is important to create a starting point where the gear and rack have no interference.

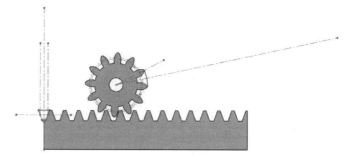

- Click the bottom linear edge of the rack in the direction of travel. Any linear edge that runs in the direction of travel is acceptable.

- Click the spur gear pitch circle as illustrated. The Pinion pitch diameter is taken from the sketch geometry. Accept the default settings.

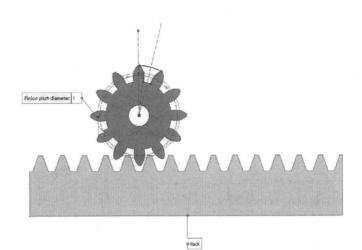

- Click OK ✔ from the RackPinionMate1 PropertyManager.

- Rotate the spur gear on the rack. View the results.

- Hide the sketches in the model.

- Close the model.

The Pitch Diameter is the diameter of an imaginary pitch circle on which a gear tooth is designed. Pitch circles of the two spur gears are required to be tangent for the gears to move and work properly.

The distance between the center point of the gear and the pitch line of the rack represents the theoretical distance required.

🔆 Diametral Pitch is the ratio equal to the number of teeth on a gear per inch of pitch diameter.

🔆 Before applying the Rack Pinion mate, make sure the teeth of the spur gear and rack are meshing properly.

Pressure Angle is the angle of direction of pressure between contacting teeth. Pressure angle determines the size of the base circle and the shape of the involute teeth. It is common for the pressure angle to be 20 or 14 1/2 degrees.

Exercise 3.19: Hinge Mate

Create a Mechanical Hinge mate. The two components in this tutorial were obtained from the SOLIDWORKS What's New section.

- Copy the Chapter 3 Homework folder to your hard drive.

- Open Hinge Mate from the Chapter 3 Homework\Hinge Mate folder. Two components are displayed.

A Hinge mate has the same effect as adding a Concentric mate plus a Coincident mate. You can also limit the angular movement between the two components.

- Click the Mate ✎ tool from the Assembly tab.

- Click the Mechanical mate tab.

- Click Hinge ⊞ mate from the Mechanical Mates box.

Set two Concentric faces.
- Click the inside cylindrical face of the first component as illustrated.

- Click the outside cylindrical face of the second component as illustrated. The two selected faces are displayed.

Set two Coincident faces.
- Click the front flat face of the first component as illustrated.

- Click the bottom flat face of the second component as illustrated. View the results in the Graphics window.

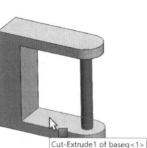

You can specify a limit angle for rotation by checking the **Specify angle limits** box and selecting the required faces.

- Click OK ✔ from the Hinge1 PropertyManager.

- Rotate the flap component about the pin.

- Close the model.

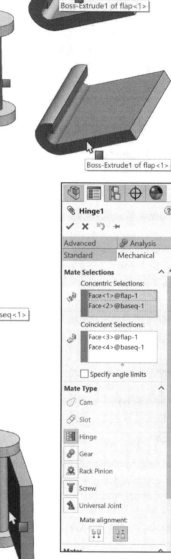

A Hinge mate limits the movement between two components to one rotational degree of freedom.

Similar to other mate types, Hinge mates do not prevent interference or collisions between components. To prevent interference, use Collision Detection or Interference Detection.

Exercise 3.20: Collision Detection

Apply the Collision Detection tool to an assembly.

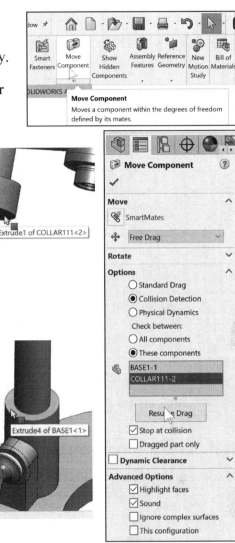

1. **Copy** the Chapter 3 Homework folder to your hard drive.

2. Open **Collision Detection** from the Chapter 3 Homework\Collision Detection folder.

3. Click the **Move Component** tool from the Assembly tab in the CommandManager. View the Move Component PropertyManager.

4. Click the **Collision Detection** box.

5. Click the **These components box**. The These components option provides the ability to select individual components for collision detection.

6. Check the **Stop at collision** box.

7. Click the **bottom face** of BASE1-1 as illustrated.

8. Click the **top face** of COLLAR111-2.

9. Click the **Resume Drag** button.

10. Click and drag the **Collar downward** towards the base as illustrated. The Collision Detection tool informs the user when there is a collision between the selected components.

11. Click **OK** ✅ from the Move Component PropertyManager.

12. **Close** the model.

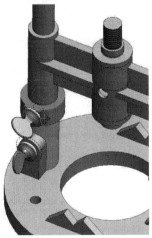

Notes:

Chapter 4

Design Modifications

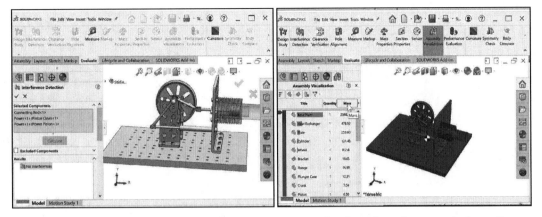

Below are the desired outcomes and usage competencies based on the completion of Chapter 4.

Desired Outcomes:	**Usage Competencies:**
• Address clearance, interference, static and dynamic behavior of the provided Stirling Engine Modified Assembly. • Verify behavior between components in the Stirling Engine Modified Assembly. • Center of Mass (COM) point for the assembly. • New Coordinate System. • A Motion Study.	• Ability to utilize the following assembly tools: Move, Rotate, Collision Detection, Interference Detection, Selected Components, Edit Feature, Center of Mass and Assembly Visualization. • Comprehend the Interference Detection tool to locate the interference between components in an assembly. • Knowledge to locate and edit the correct mates. • Apply the Measure and Mass Properties tool. • Create a new Coordinate System relative to the origin of the assembly. • Apply motion to an assembly.

Notes:

Chapter 4 - Design Modifications

Chapter Objective

Before you machine or create a rapid prototype of a part for an assembly, verify clearance, interference, static and dynamic behavior between the assembly parts. Apply the Measure tool to fix a distance mate.

Calculate properties on the parts in an assembly: mass, density, and volume based on the model geometry and material properties. Add a Center of Mass (COM) point to an assembly. The COM point updates when the model's center of mass changes. Create a new Coordinate system for an assembly. Apply motion to an assembly.

On the completion of this chapter, you will be able to:

- Establish a SOLIDWORKS session.

- Utilize the following assembly tools: Move, Rotate, Collision Detection, Interference Detection, Selected Components, Edit Feature, Edit Mate, Center of Mass and Assembly Visualization.

- Edit a Distance mate.

- Insert a Center of Mass (COM) point into the Assembly.

- Create a new Coordinate System.

- Apply the Measure and Mass Properties tool.

- Utilize the Assembly Visualization tool on the Stirling Engine Modified Assembly and sort components by mass.

- Apply motion to an assembly.

Activity: Start a SOLIDWORKS Session.

Start a SOLIDWORKS session.
1) Double-click the **SOLIDWORKS icon** from the desktop.

2) If needed, **close** the SOLIDWORKS Welcome dialog box.

In the next section, open the Stirling Engine Modified assembly from the SOLIDWORKS 2024\Design Modifications folder.

Verify for clearance, interference, static and dynamic behavior between the Power Piston, Power Clevis, Connecting Rod and Handle.

Activity: Open the Stirling Engine Modified Assembly.

Open the Stirling Engine Modified Assembly.

3) Click **File**, **Open** from the Main menu.

4) **Browse** to the SOLIDWORKS 2024\Design Modifications folder.

5) Double-click the **Stirling Engine Modified** assembly. The assembly is displayed in the Graphics window.

6) **Rotate** the Handle to view the rotation of the assembly.

7) **Position** the Handle at the top as illustrated.

Verify for clearance, interference, static and dynamic behavior between the Power Piston, Power Clevis, Connecting Rod and Handle in the Stirling Engine Modified Assembly.

Detect collisions with other components when moving or rotating a component. Detect collisions with the entire assembly or a selected group of components.

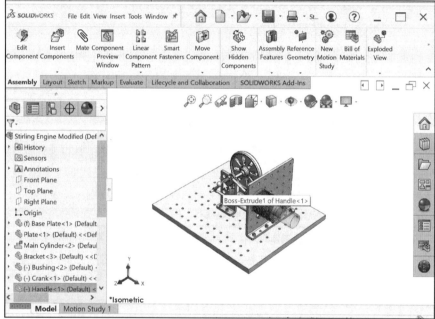

Physical Dynamics is an option in Collision Detection that allows you to view the motion of assembly components in a realistic way. With Physical Dynamics enabled, when you drag a component, the component applies a force to components that it touches and moves the components if they are free to translate or rotate.

| **Activity: Verify Collision between Components in the Assembly.** |

Activate the Collision Detection tool.

8)   Click the **Assembly** tab from the
CommandManager.

9)   Click the **Move Component** 🔲 icon.
The Move Component PropertyManager
is displayed.

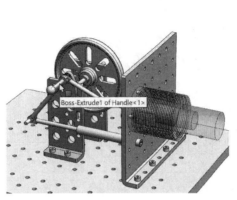

10) Click the **Collision
Detection** button.

11) Check the **Highlight
faces**, **Sound** and
**Ignore complex
surfaces** box from
the Advanced
Options section.
Note: Increase the
volume on your
computer.

12) Uncheck the **Stop at
collision** box.

13) Click the **These
components** button. If
needed right-click inside
the Components for
Collision Check box and
click **Clear Selections**.

14) Click the **Handle** from
the Graphics window.
Handle-1 is displayed in
the Components for
collision Check box.

15) Click the **Connecting
Rod** from the Graphics
window. Connection
Rod-1 is displayed in
the Components for Collision Check box.

16) **Rotate** and **Zoom-in** on the Piston Clevis.

17) Click the **Piston Clevis** as illustrated from the
Graphics window. Power-1/Piston Clevis 1 is
displayed in the Components for collision Check
box.

18) **Display** a Front view.

19) Click the **Power Piston** as illustrated from the Graphics window. Power-1/Power Piston-1 is displayed in the Components for Collision Check box.

20) **Display** a Trimetric view.

21) Click the **Resume Drag** button.

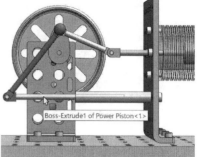

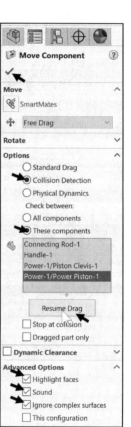

22) Slowly drag the **Handle collar (Boss-Extrude1 of Collar <1>)** downward and stop when the components collide. Collision components are displayed in the Graphics window and a sound is made. There are collisions between the Connecting Rod and the Power Piston and the Connecting Rod and the Piston Clevis.

23) Click **OK** from the Move Component PropertyManager.

In the next section, utilize the Interference Detection tool to locate the interference between components.

Interferece detection is useful in complex assemblies, where it can be difficult to visually determine whether components interfere with each other.

Activity: Utilize the Interference Detection tool.

Activate the Interference Detection tool.

24) Click the **Evaluate** tab from the CommandManager.

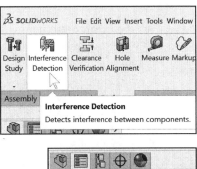

25) Click the **Interference Detection** icon. The Interference Detection PropertyManager is displayed.

26) Right-click in the **Selected Components** box.

27) Click **Clear Selections**. By default, the top-level assembly appears unless you preselect other components. When you check an assembly for interference, all of its components are checked. If you select a single component, only the interferences that involve that component are reported. If you select two or more components, only the interferences between the selected components are reported.

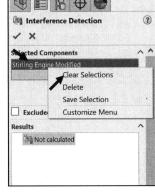

Select component to perform the Interference detection process.

28) Click the **Connecting Rod** from the Graphics window as illustrated. Connection Rod-1 is displayed in the Selected Components box.

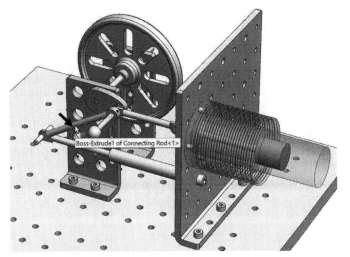

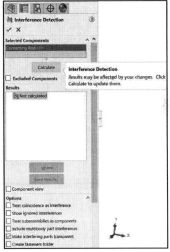

29) **Rotate** the model to view the Piston Clevis.

30) Click the **Piston Clevis** from the Graphics window as illustrated.

31) Click the **Power Piston** from the Graphics window as illustrated.

32) Click the **Calculate** button. When you select an interference under Results, it highlights in red in the Graphics window. *The volume interference value and the number of interferences will be different depending* on the position of the Crank in step 17.

33) Click each **Interference Results** to view the interference between the components displayed in red.

34) Click **OK** ✔ from the Interference Detection PropertyManager.

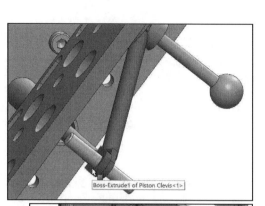

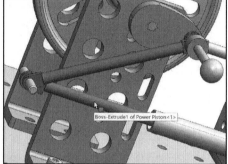

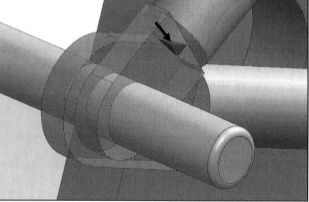

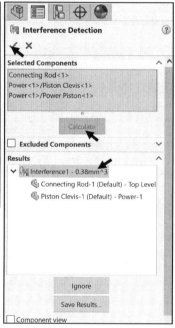

The Interference Detection and Collision Detection results require design changes to the Power Piston, Piston Clevis, Connecting Rod and other components in the Stirling Engine Modified Assembly. Review Assembly and Evaluate tools to modify the design intent.

Activity: Modify the Connecting Distance2 Rod Mate.

View the Mates on the Connecting Rod in the Assembly.

35) **Expand** Connecting Rod in the FeatureManager.

36) **Expand** the Mates in Stirling Engine Modified folder as illustrated.

37) Click on each **Mate** to review how the Connecting Rod is constrained in the Stirling Engine Modified assembly.

Modify the Distance Mate between the face of the Rod and the face of the Connection Rod.

38) Right-click **Distance2** from the FeatureManager.

39) Click the **Edit Feature** 🔧 icon from the Context toolbar. The current distance is 0mm.

40) Enter **3mm** for distance.

41) Click **OK** ✔ from the Distance2 PropertyManager.

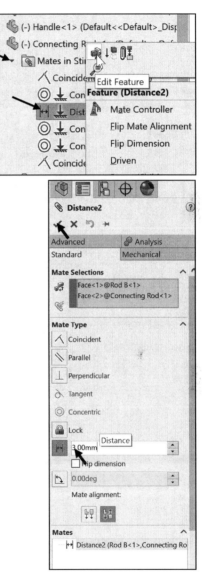

Display a Trimetric view.

42) Click **Trimetric view** 🔲 from the Heads-up toolbar.

Save the model.

43) Click **Save** 💾.

44) Click **Rebuild and Save the document**.

45) Click **Save All**.

💡 A Distance Mate of 0mm allows flexibility when you need to vary spacing between components.

💡 If you delete a Mate and then recreate it, the Mate numbers will be different (increase).

Utilize the Measure tool to check the Distance mate in the Graphics window.

46) Click the **Evaluate** tab in the CommandManager.

47) Click the **Measure** tool. The Measure dialog box is displayed. Expand the Measure dialog box if needed.

48) Click the **Show XYZ Measurements** button.

49) Right-click in the **Selection** box.

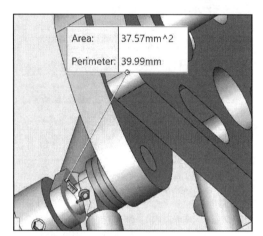

50) Click **Clear Selections**.

51) **Zoom in** on the two faces (Rod and Connecting Rod).

52) Click the **Rod face** as illustrated.

53) Click the **Connecting Rod face**. The Normal Distance is 3.00mm.

Close the Measure Dialog box.

54) Click **Close** ⊠.

Display a Trimetric view.

55) Click **Trimetric view** from the Heads-up toolbar.

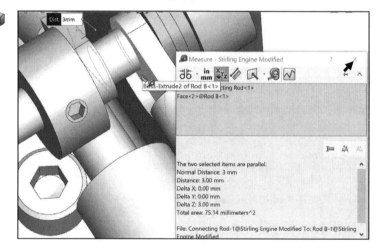

Save the model.

56) Click **Save** 💾.

Activity: Utilize the Interference Detection tool - Check the Solution.

Activate the Interference Detection tool.

57) Click the **Evaluate** tab from the CommandManager.

58) Click the **Interference Detection** icon. The Interference Detection PropertyManager is displayed.

59) Right-click in the **Selected Components** box.

60) Click **Clear Selections**.

Select component to perform the Interference detection process.

61) Click the **Connecting Rod** from the Graphics window as illustrated. Connection Rod-1 is displayed in the Selected Components box.

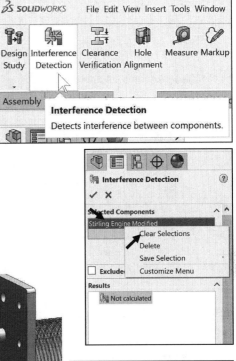

62) **Rotate** the model to view the Piston Clevis.

63) Click the **Piston Clevis** from the Graphics window as illustrated.

64) Click the **Power Piston** from the Graphics window as illustrated.

65) Click the **Calculate** button. There are no interferences between the components. The modification of the Distance mate resolved the interference issue.

66) Click **OK** ✔ from the Interference Detection PropertyManager.

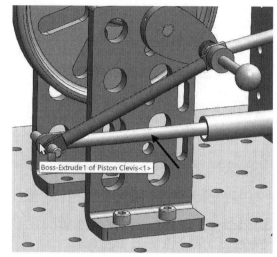

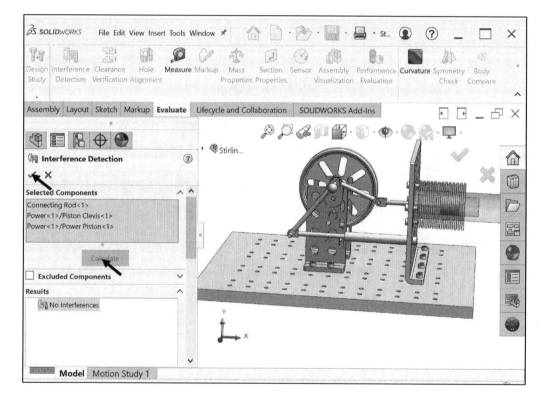

Activity: Locate the Center of Mass of the Assembly.

Locate the Center of Mass along with other Mass Properties of the assembly (COM). You can assign values for mass, center of mass and moments of inertia to override the calculated values.

View the Mass Properties of the Assembly.

67) Click the **Evaluate** tab from the CommandManager.

68) Click the **Mass Properties** icon. The Mass Properties dialog box is displayed. View the results. The total mass is 3447.14 grams. The numbers represent the document properties (2 decimal places). Your Center of mass will be different due to the location of the Handle in the assembly.

View the Mass Properties of the Plate in the Assembly.

69) Right-click inside the **Selection** box.

70) Click **Clear Selections**.

71) Click the **Plate** in the Graphics window as illustrated. Plate-1 is displayed in the Selection box. The Mass is approximately 259.60 grams.

Return to the Mass Properties of the Assembly.

72) Right-click inside the **Selection** box.

73) Click **Clear Selections**.

74) Click **Stirling Engine Modified** from the Assembly FeatureManager. Stirling Engine Modified is displayed in the Mass Properties Selection box.

Locate the Center of Mass of the Assembly.

75) Click the **Create Center of Mass feature** box.

Close the Mass Properties dialog box.

76) Click **Close** ✕.

Display the COM point in the Assembly.

77) If needed, click **View Center of Mass** from the Hide/Show Items in the Heads-up toolbar.

78) Click a **position** in the Graphics window. View the results.

Display a Trimetric - Shaded With Edges view.

79) Press the **space bar** to display the Orientation dialog box.

80) Click **Trimetric** ⬜. You can also access the Isometric view tool from the Heads-up View toolbar.

81) Click the **Shaded With Edges** ⬜ icon.

Save the Assembly.

82) Click **Save** 🖫.

83) Click **Rebuild** and **save**.

You can add a Center of Mass (COM) point to parts and assemblies. In drawings of parts or assemblies that contain a COM point, you can show and reference the COM point.

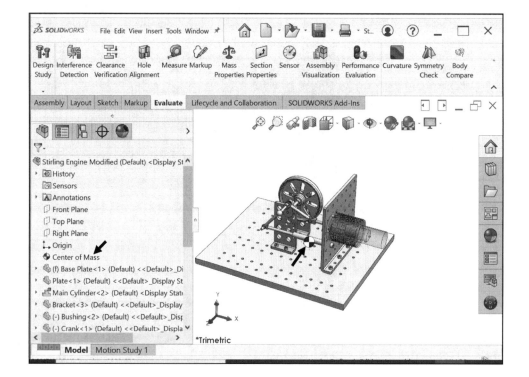

You can define a coordinate system for a part or assembly. Coordinate systems are useful:

- With the Measure and Mass Properties tools.

- When exporting SOLIDWORKS documents to IGES, STL, ACIS, STEP, Parasolid, VRML and VDA.

- When applying assembly mates.

In the next section, create a new coordinate system on the front bottom vertex of the assembly.

Activity: Create a new Coordinate System in the Assembly.

View the new Coordinate System on the front bottom vertex of the assembly.

84) Click the **Assembly** tab from the CommandManager.

85) Click **Coordinate System** ⅃ from the Consolidated Reference Geometry drop-down menu. The Coordinate System PropertyManager is displayed.

86) Click the **bottom front vertex** of the Assembly Base Plate. Vertex<1> is displayed.

87) Click **OK** ✔ from the Coordinate System PropertyManager. Coordinate System1 is displayed in the Assembly FeatureManager.

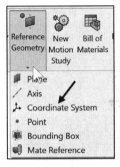

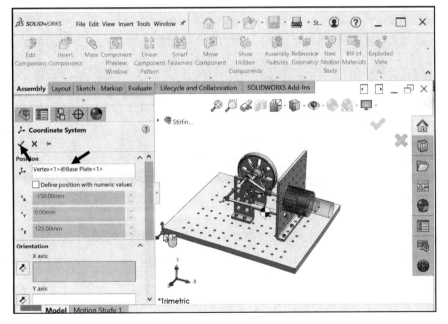

In the next section, display the Mass Properties relative to the new Coordinate system of the assembly.

> **Activity: Display the Mass Properties relative to the new Coordinate System.**

View the new Center of Mass point.

88) Click the **Evaluate** tab from the CommandManager.

89) Click the **Mass Properties** ⚖ icon. The Mass Properties dialog box is displayed.

90) Click the **drop-down arrow** for the Report coordinate values relative to:

91) Select **Coordinate System1**. View the updated results (Center of Mass) in the Mass Properties dialog box. Your Center of mass will be different due to the original location of the Handle in the assembly.

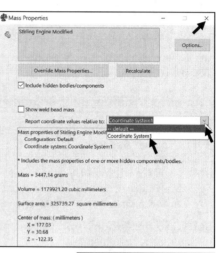

Close the Mass Properties dialog box.

92) Click **Close** ☒.

Display a Trimetric view.

93) Press the **space bar** to display the Orientation dialog box.

94) Click **Trimetric view** 📦. You can also access the Isometric view tool from the Heads-up View toolbar.

Assembly Visualization

In an assembly, the designer selects material based on cost, performance, physical properties, manufacturing processes, sustainability, etc.

The SOLIDWORKS Assembly Visualization tool includes a set of predefined columns to help troubleshoot assembly performance. You can view the open and rebuild times for the components, and the total number of graphics triangles for all instances of components.

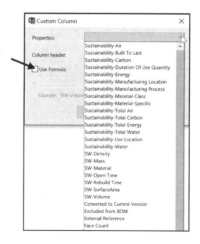

The Assembly Visualization tool provides the ability to rank components based on the default values (**weight, mass, density, volume, etc.**) or their custom properties (**cost, sustainability, density, surface area, volume, etc.**) or an equation and activate a spectrum of colors that reflects the relative values of the properties for each component.

Hide/Show Value Bar ▤ **icon**. Available for numeric properties. Turns the value bars off and on. When the value bars are on, the component with the highest value displays the longest bar. You can set the length of the bars to be calculated relative to the highest-value component or relative to the entire assembly.

Flat Nested View 🔲 **icon**. Nested view, where subassemblies are indented. Flat view, where subassembly structures are ignored (similar to a parts-only BOM).

Grouped/Ungrouped View 🔲 **icon**. Groups multiple instances of a component into a single line item in the list. Grouped View is useful when listing values for properties that are identical for every instance of the component. Ungrouped view lists each instance of a component individually. Ungrouped View is useful when listing values for instance-specific properties, such as Fully mated, which might be different for different instances of the component.

Performance Analysis 🔲 **icon**.
Provides additional information on the open, display, and rebuild performance of models in an assembly.

Filter ▽ ˙ **icon**. Filters the list by text and by component show/hide state.

The Assembly Visualization 🔲 tab in the FeatureManager design tree panel contains a list of all components in the assembly, sorted initially by file name. There are three default columns:

- File name

- Quantity

- Mass

	Title	Quantity	Mass		
				Total Weight	
	2MM Set Screw	2	0.01	Mass	
	2MM Set Screw...	1	0.02	Density	
	Piston Clevis	1	0.15	Volume	
	B18.3.1M - 4 x 0...	4	0.26	Surface Area	
	Spacer	3	0.26	More...	
	Collar(Default)	3	0.32	Add Column	
	Collar(Collar 4...	1	0.32	Unit Precision	
	B18.3.1M - 4 x 0...	6	0.33	Value Bars	
	B18.3.1M - 5 x 0...	1	0.43	Add Display State	
	4MM Pin Short	1	0.97	Load Style	
	4MM Pin Long	1	2.44	Save Style	
				Save As ...	
				Performance Analysis	

💡 Mass is the default column. View your options as illustrated other than Mass.

You can save the list information in a separate file such as a Microsoft® Excel® spreadsheet or a text file.

In the next section, utilize the Assembly Visualization tool on the Stirling Engine Modify Assembly. Sort the assembly components by mass.

Activity: Assembly Visualization tool - Sort Components by Mass.

Apply the Assembly Visualization tool. Sort by mass.

95) Click the **Evaluate** tab from the CommandManager.

96) Click the **Assembly Visualization**  icon. The Assembly Visualization PropertyManager is displayed.

97) Click the **arrow** to the right of the File Name row as illustrated.

98) Select **Mass**.

99) Click the **Mass** column.

100) View the results from low to high or from high to low.

101) Click the **Toggle Slider** to view the color of the components in the Graphics window.

102) Explore the other **columns** and **options**.

103) View the results in the Graphics window.

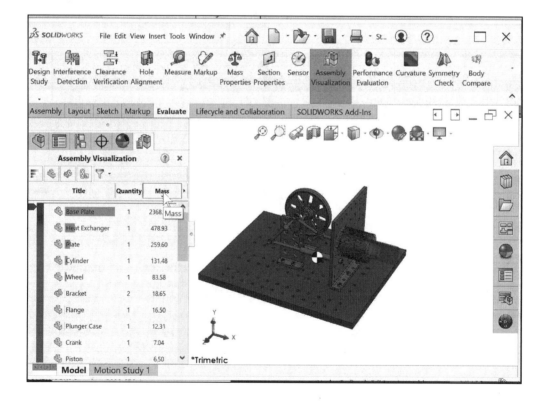

Return to the Assembly FeatureManager.

104) Click the Assembly **FeatureManager Design tree** 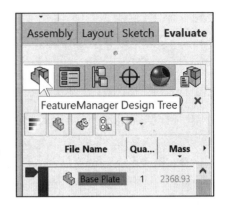 icon.

Display a Trimetric view.

105) Press the **space bar** to display the Orientation dialog box.

106) Click **Trimetric view** . You can also access the Isometric view tool from the Heads-up View toolbar.

Save the Assembly.

107) Click **Save** .

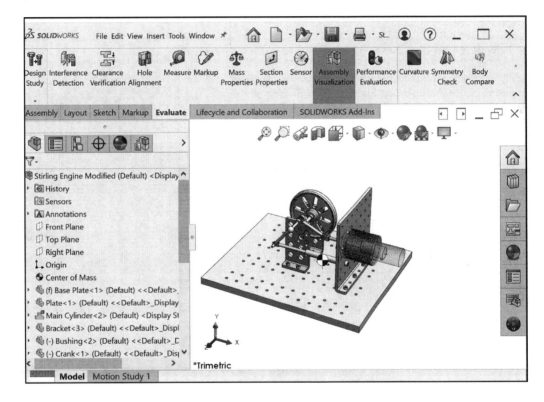

Motion Study

Motion studies are graphical simulations of motion for assembly models. You can incorporate visual properties such as lighting and camera perspective into a motion study.

Motion studies do not change an assembly model or its properties. They simulate and animate the motion you prescribe for a model. You can use SOLIDWORKS mates to restrict the motion of components in an assembly when you model motion.

In the next section, play the Basic Motion study of the final assembly. You can use Basic Motion for approximating the effects of motors, springs, contact and gravity on assemblies. Basic Motion takes mass into account in calculating motion. Basic Motion computation is relatively fast, so you can use this for creating presentation-worthy animations using physics-based simulations. A simple rotary motor was applied to the Handle for movement.

Activity: Play the Basic Motion Study of the Assembly.

Play the Basic Motion Study. Access the Motion Manager from the Motion Study tab.

108) Click the **Motion Study #** tab at the bottom left of the Graphics window. The Motion Manager is displayed.

| | **Model** | Motion Study 1 |

109) Select **Animation** from the Motion Study drop-down menu. A motion study was created using a simple Rotary Motor on the Handle.

110) Click the **Calculate** icon.

111) Click the **Play from Start** . View the Motion Study in the Graphics window.

Create and save a Motion Study.

112) Click the **Save Animation** icon. View the Save Animation to File dialog box.

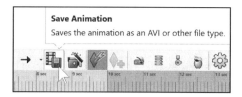

Save Animation
Saves the animation as an AVI or other file type.

113) Select a **Save in** location. View your options.

114) Click **Save**. View your options.

115) Select **MP4 video file (.mp4)**.

116) Click **OK** from the Video Compression dialog box.

Return to the Model.

117) Click the **Model** tab at the bottom left of the Graphics window.

118) **Explore** SOLIDWORKS Help for additional information on Basic Motion using timers and other tools.

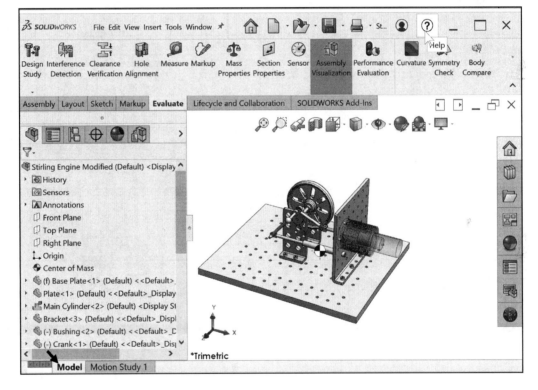

Save the Assembly.

119) Click **Save** 💾.

120) Click **Rebuild** and **save**.

Close all models.

121) Click **Window**, **Close All** from the Main menu bar.

You are finished with this section.

Summary

In this chapter, you addressed clearance, interference, static and dynamic behavior of the Stirling Engine Modified assembly.

You verified the behavior between the following components: Power Piston, Power Clevis, Connecting Rod and Handle in the assembly.

The following assembly tools were utilized in the chapter: Move, Rotate, Collision Detection, Interference Detection, Selected Components, Edit Feature and Center of Mass.

The Assembly Visualization tool was utilized on the Stirling Engine assembly and sorted by component mass.

A new Coordinate System was created on the Stirling Engine assembly relative to the default origin.

A Motion Study was created and saved.

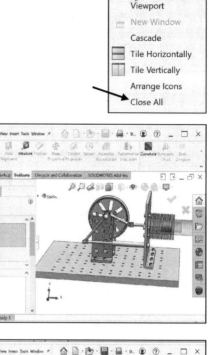

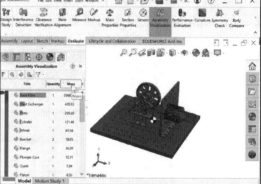

In Chapter 5, you will learn about the Drawing and Dimension Fundamentals and create two new drawings with user defined document properties and custom properties:

- Fly Wheel Assembly.

- Bushing.

Create the Fly Wheel Assembly drawing with an Exploded Isometric view.

Utilize a Bill of Materials and Balloons.

Learn about Custom Properties and the Title Block.

Create the Bushing Part drawing utilizing Third Angle Projection with two standard Orthographic views: Front, Top and an Isometric view.

Address imported dimensions from the Model Items tool.

Insert additional dimensions using the Smart Dimension tool along with all needed annotations.

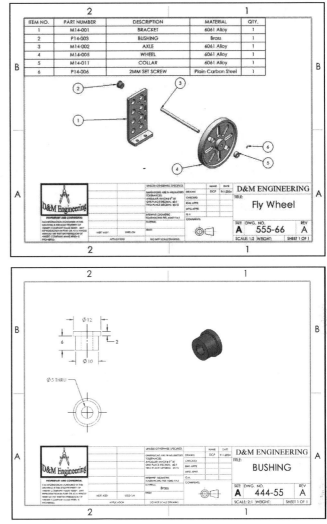

Questions

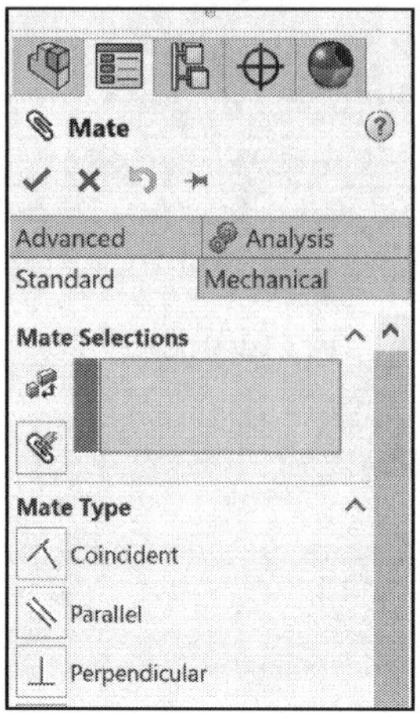

1. True or False. Mates provide the ability to create geometric relationships between assembly components. Mates define the allowable directions of rotational or linear motion of the components in the assembly.

2. True or False. If you delete a mate and then recreate it, the mate number will be in a different order (increase).

3. True or False. To fix the first component to the Origin in an assembly, click OK ✔ from the Begin Assembly PropertyManager or click the Origin in the Graphics window.

4. Describe an Angle mate in SOLIDWORKS.

5. Describe a Coincident mate in SOLIDWORKS.

6. Describe an assembly or sub-assembly in SOLIDWORKS.

7. True or False. To remove the fixed state of a component in an assembly, Right-click the component name in the FeatureManager. Click Float. The component is free to move.

8. In an assembly, each component has_____ # degrees of freedom. Name them.

9. True or False. Mates are geometric relationships that align and fit components in an assembly. Mates remove degrees of freedom from a component.

10. List two Mechanical mate types.

11. List two Advanced mate types.

12. List two Standard mate types.

13. True or False. A Part Template is the foundation for an SOLIDWORKS assembly document.

14. True or False. Determine the static and dynamic behavior of mates in each sub-assembly before creating the top-level assembly.

Exercises

Exercise 4.1: Butterfly Valve Project

Use the web to locate a similar assembly and to understand the parts needed in the assembly. View the provided sample parts. Create your own parts and final assembly. View the provided motion file to create the proper movement in SOLIDWORKS.

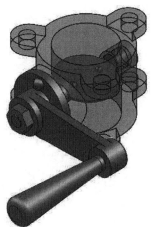

Create an ANSI Butterfly valve assembly document. Insert all needed components and mates to assemble the assembly. Create and insert any additional components if needed.

Create a A-ANSI Landscape - Third Angle Isometric Exploded Drawing document with Explode lines of the assembly using your knowledge of SOLIDWORKS. Insert a BOM with Balloons. Insert all needed General notes and Custom Properties in the Title Block.

- Arbor Press Project
- Bench Vice Project
- Butterfly Valve Project
- Drill Guide Project
- Kant Twist Clamp Project
- Pipe Vice Project
- Pulley Project
- Quick Acting Clamp Project
- Radial Engine Project
- Shaper Tool Head Project
- Welder Arm Project

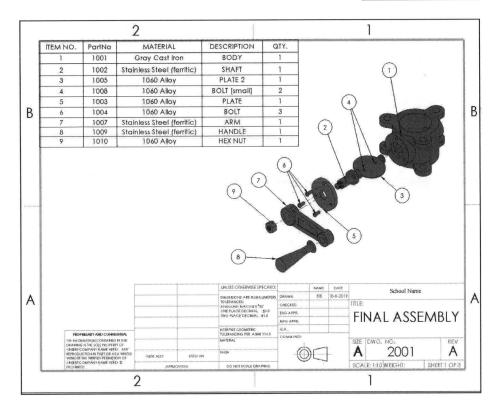

ITEM NO.	PartNo	MATERIAL	DESCRIPTION	QTY.
1	1001	Gray Cast Iron	BODY	1
2	1002	Stainless Steel (ferritic)	SHAFT	1
3	1005	1060 Alloy	PLATE 2	1
4	1008	1060 Alloy	BOLT [small]	2
5	1003	1060 Alloy	PLATE	1
6	1004	1060 Alloy	BOLT	3
7	1007	Stainless Steel (ferritic)	ARM	1
8	1009	Stainless Steel (ferritic)	HANDLE	1
9	1010	1060 Alloy	HEX NUT	1

TITLE: FINAL ASSEMBLY

SIZE A   DWG. NO. 2001   REV A

Exercise 4.2: Quick Acting Clamp Project

Use the web to locate a similar assembly and to
understand the parts needed in the assembly. View
the provided sample parts. Create your own parts
and final assembly. View the provided motion file
to create the proper movement in SOLIDWORKS.

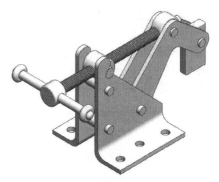

Create an ANSI Quick Acting clamp assembly
document. Insert all needed components and mates
to assemble the assembly. Create and insert any
additional components if needed.

Create a C-ANSI Landscape - Third Angle Isometric Exploded
Drawing document with Explode lines of the assembly using your
knowledge of SOLIDWORKS. Insert a BOM with Balloons. Insert
all needed General notes and Custom Properties in the Title Block.

Arbor Press Project
Bench Vice Project
Butterfly Valve Project
Drill Guide Project
Kant Twist Clamp Project
Pipe Vice Project
Pulley Project
Quick Acting Clamp Project
Radial Engine Project
Shaper Tool Head Project
Welder Arm Project

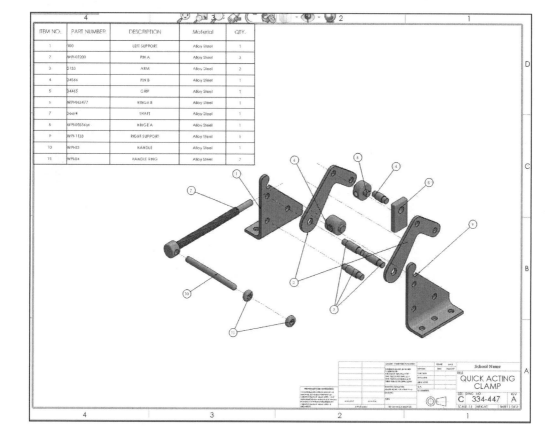

ITEM NO.	PART NUMBER	DESCRIPTION	Material	QTY.
1	100	LEFT SUPPORT	Alloy Steel	1
2	WPI-07200	PIN A	Alloy Steel	2
3	2133	ARM	Alloy Steel	2
4	34566	PIN B	Alloy Steel	1
5	34465	GRIP	Alloy Steel	1
6	WPI-063477	HINGE B	Alloy Steel	1
7	0ertt4	SHAFT	Alloy Steel	1
8	WPI-05556ye	HINGE A	Alloy Steel	1
9	WPI-1133	RIGHT SUPPORT	Alloy Steel	1
10	WPI-03	HANDLE	Alloy Steel	1
11	WPI-04	HANDLE RING	Alloy Steel	2

Exercise 4.3: Drill Guide Project

Use the web to locate a similar assembly and to understand the parts needed in the assembly. View the provided sample parts. Create your own parts and final assembly. View the provided motion file to create the proper movement in SOLIDWORKS.

Create an ANSI Drill Guide assembly document. Insert all needed components and mates to assemble the assembly. Create and insert any additional components if needed.

Create a C-ANSI Landscape - Third Angle Isometric Exploded Drawing document with Explode lines of the assembly using your knowledge of SOLIDWORKS. Insert a BOM with Balloons. Insert all needed General notes and Custom Properties in the Title Block.

Arbor Press Project
Bench Vice Project
Butterfly Valve Project
Drill Guide Project
Kant Twist Clamp Project
Pipe Vice Project
Pulley Project
Quick Acting Clamp Project
Radial Engine Project
Shaper Tool Head Project
Welder Arm Project

ITEM NO.	PART NUMBER	DESCRIPTION	MATERIAL	QTY.
1	WPI-1000-01	BASE	2014 ALLOY	1
2	WPI-1000-02	ROTATOR	PLAIN CARBON STEEL	2
3	WPI-1000-03	ROD GUIDE	CAST ALLOY STEEL	2
4	WPI-1000-04	SLIDE	2014 ALLOY	1
5	WPI-1000-05	COLLAR	2014 ALLOY	2
6	WPI-1000-06	BUSHING	ALUMINUM BRONZE	2
7	WPI-1000-07	RETAINING RING B27.1 - NA2-45	ALLOY STEEL	2
8	WPI-1000-08	DRILL ADAPTOR	PLAIN CARBON STEEL	1
9	WPI-1000-09	THUMB SCREW .25-20x0.51 TYPE B, FLAT POINT-C	2014 ALLOY	6

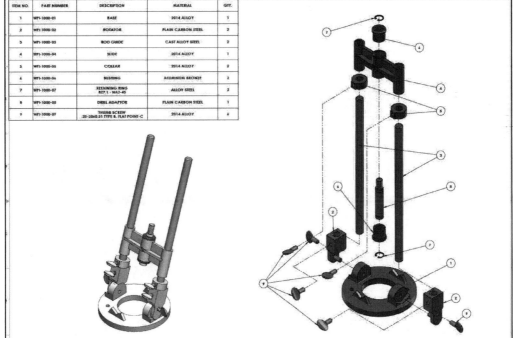

Exercise 4.4: Pulley Project

Use the web to locate a similar assembly and to understand the parts needed in the assembly. View the provided sample parts. Create your own parts and final assembly. View the provided motion file to create the proper movement in SOLIDWORKS.

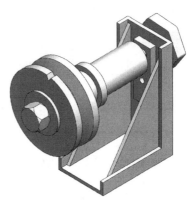

Create an ANSI Pully assembly document. Insert all needed components and mates to assemble the assembly. Create and insert any additional components if needed.

Create a A-ANSI Landscape - Third Angle Isometric Exploded Drawing document with Explode lines of the assembly using your knowledge of SOLIDWORKS. Insert a BOM with Balloons. Insert all needed General notes and Custom Properties in the Title Block.

Note: Additional components were added below.

Arbor Press Project
Bench Vice Project
Butterfly Valve Project
Drill Guide Project
Kant Twist Clamp Project
Pipe Vice Project
Pulley Project
Quick Acting Clamp Project
Radial Engine Project
Shaper Tool Head Project
Welder Arm Project

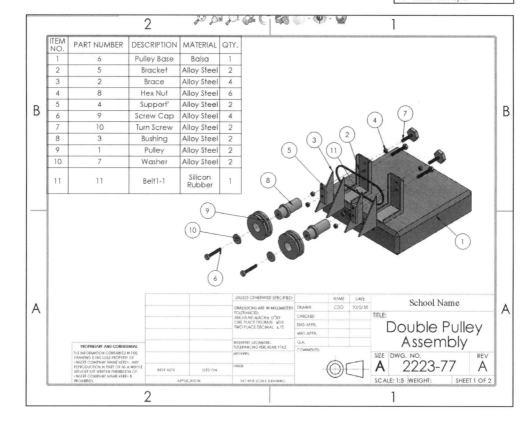

ITEM NO.	PART NUMBER	DESCRIPTION	MATERIAL	QTY.
1	6	Pulley Base	Balsa	1
2	5	Bracket	Alloy Steel	2
3	2	Brace	Alloy Steel	4
4	8	Hex Nut	Alloy Steel	6
5	4	Support'	Alloy Steel	2
6	9	Screw Cap	Alloy Steel	4
7	10	Turn Screw	Alloy Steel	2
8	3	Bushing	Alloy Steel	2
9	1	Pulley	Alloy Steel	2
10	7	Washer	Alloy Steel	2
11	11	Belt1-1	Silicon Rubber	1

UNLESS OTHERWISE SPECIFIED:
DIMENSIONS ARE IN MILLIMETERS
TOLERANCES:
ANGULAR: MACH ± 0°30'
ONE PLACE DECIMAL ±0.5
TWO PLACE DECIMAL ±.15

INTERPRET GEOMETRIC
TOLERANCING PER ASME Y14.5
MATERIAL

FINISH

PROPRIETARY AND CONFIDENTIAL
THE INFORMATION CONTAINED IN THIS DRAWING IS THE SOLE PROPERTY OF <INSERT COMPANY NAME HERE>. ANY REPRODUCTION IN PART OR AS A WHOLE WITHOUT THE WRITTEN PERMISSION OF <INSERT COMPANY NAME HERE> IS PROHIBITED.

NEXT ASSY   USED ON
APPLICATION   DO NOT SCALE DRAWING

DRAWN   CSG   10/5/18
CHECKED
ENG APPR.
MFG APPR.
Q.A.
COMMENTS:

School Name

TITLE:
Double Pulley Assembly

SIZE DWG. NO.   REV
A   2223-77   A

SCALE: 1:5 WEIGHT: SHEET 1 OF 2

Exercise 4.5: Welder Arm Project

Use the web to locate a similar assembly and to understand the parts needed in the assembly. View the provided sample parts. Create your own parts and final assembly. View the provided motion file to create the proper movement in SOLIDWORKS.

Create an ANSI Welder Arm assembly document. Insert all needed components and mates to assemble the assembly. Create and insert any additional components if needed.

Create a C-ANSI Landscape - Third Angle Isometric Exploded Drawing document with Explode lines of the assembly using your knowledge of SOLIDWORKS. Insert a BOM with Balloons. Insert all needed General notes and Custom Properties in the Title Block.

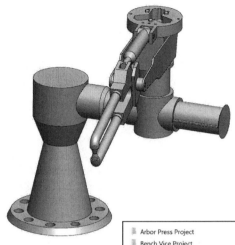

Arbor Press Project
Bench Vice Project
Butterfly Valve Project
Drill Guide Project
Kant Twist Clamp Project
Pipe Vice Project
Pulley Project
Quick Acting Clamp Project
Radial Engine Project
Shaper Tool Head Project
Welder Arm Project

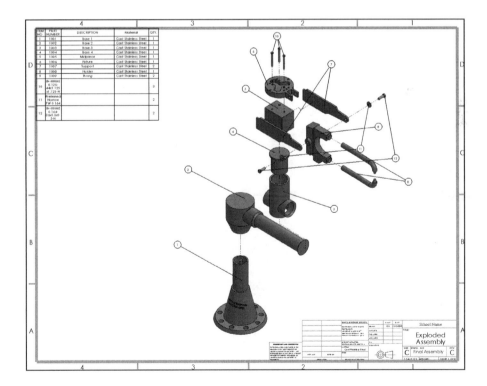

Exercise 4.6: Radial Engine Project

Use the web to locate a similar assembly and to understand the parts needed in the assembly. View the provided sample parts. Create your own parts and final assembly. View the provided motion file to create the proper movement in SOLIDWORKS.

Create an ANSI Radial Engine assembly document. Insert all needed components and mates to assemble the assembly. Create and insert any additional components if needed.

Create a C-ANSI Landscape - Third Angle Isometric Exploded Drawing document with Explode lines of the assembly using your knowledge of SOLIDWORKS. Insert a BOM with Balloons. Insert all needed General notes and Custom Properties in the Title Block.

Note: Additional components were added to the drawing.

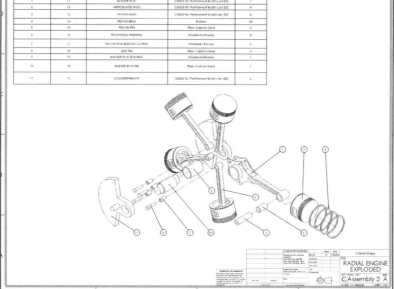

ITEM NO.	PART NUMBER	DESCRIPTION	MATERIAL	QTY.
1	11	MASTER ROD	C355.0-T61 Permanent Mold cast (SS)	1
2	12	ARTICULATED ROD	C355.0-T61 Permanent Mold cast (SS)	4
3	13	PISTON HEAD	C355.0-T61 Permanent Mold cast (SS)	5
4	14	PISTON RING	Rubber	20
5	15	PISTON PIN	Plain Carbon Steel	5
6	16	PISTON ROD BUSHING	Aluminum Bronze	5
7	17	PISTON ROD BUSHING LOWER	Aluminum Bronze	4
8	18	LINK PIN	Plain Carbon Steel	4
9	19	MASTER ROD BUSHING	Aluminum Bronze	1
10	10	MASTER ROD PIN	Plain Carbon Steel	1
11	11	COUNTERWEIGHT	C355.0-T61 Permanent Mold cast (SS)	2

RADIAL ENGINE EXPLODED

C Assembly 2 A

Exercise 4.7: Bench Vice Project

Use the web to locate a similar assembly and to understand the parts needed in the assembly. View the provided sample parts. Create your own parts and final assembly. View the provided motion file to create the proper movement in SOLIDWORKS.

Create an ANSI Bench Vice assembly document. Insert all needed components and mates to assemble the assembly. Create and insert any additional components if needed.

Create an A-ANSI Landscape - Third Angle Isometric Exploded Drawing document with Explode lines of the assembly using your knowledge of SOLIDWORKS.

Insert a BOM with Balloons. Insert all needed General notes and Custom Properties in the Title Block.

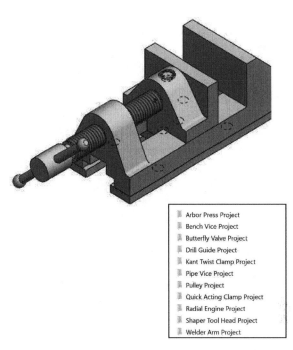

Arbor Press Project
Bench Vice Project
Butterfly Valve Project
Drill Guide Project
Kant Twist Clamp Project
Pipe Vice Project
Pulley Project
Quick Acting Clamp Project
Radial Engine Project
Shaper Tool Head Project
Welder Arm Project

ITEM NO.	PART NUMBER	DESCRIPTION	MATERIAL	QTY.
1	WPI-0001	Base	Alloy Steel	1
2	WPI-0002	Base Plate	Alloy Steel	2
3	WPI-0003	Vice Jaw	Alloy Steel	1
4	WPI-0004	Clamping Plate	Alloy Steel	1
5	WPI-0005	Jaw Screw	Alloy Steel	1
6	WPI-0006	Screw Bar	Alloy Steel	1
7	WPI-0007	Bar Globes	Alloy Steel	2
8	WPI-008	Hex Screw M6x1.0x14-14WN	Alloy Steel	4
9	WPI-0009	Bolt	Alloy Steell	1
10	WPI-00019	Blotl	Alloy Steel	2

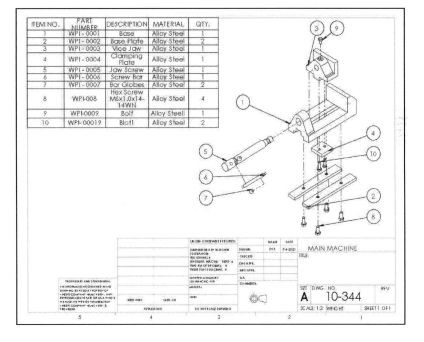

Exercise 4.8: Kant Twist Clamp Project

Use the web to locate a similar assembly and to understand the parts needed in the assembly. View the provided sample parts. Create your own parts and final assembly. View the provided motion file to create the proper movement in SOLIDWORKS.

Create an ANSI Kant Twist clamp assembly document. Insert all needed components and mates to assemble the assembly. Create and insert any additional components if needed.

Create a C-ANSI Landscape - Third Angle Isometric Exploded Drawing document with Explode lines of the assembly using your knowledge of SOLIDWORKS. Insert a BOM with Balloons. Insert all needed General notes and Custom Properties in the Title Block.

Note: Additional components were added to the drawing.

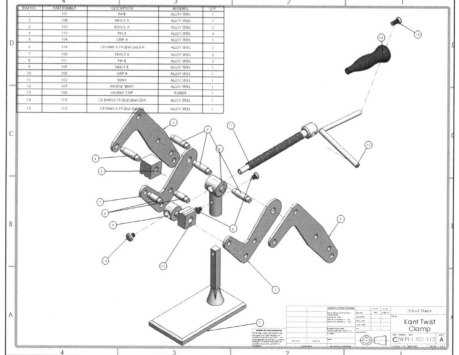

ITEM NO.	PART NUMBER	DESCRIPTION	MATERIAL	QTY
1	101	BASE	ALLOY STEEL	1
2	108	HINGE A	ALLOY STEEL	1
3	102	BRACE A	ALLOY STEEL	2
4	110	PIN A	ALLOY STEEL	4
5	104	GRIP A	ALLOY STEEL	1
6	114	CR-FHMS 0.19-32x0.5x0.5-N	ALLOY STEEL	2
7	103	BRACE B	ALLOY STEEL	2
8	111	PIN B	ALLOY STEEL	2
9	109	HINGE B	ALLOY STEEL	1
10	105	GRIP B	ALLOY STEEL	1
11	112	SHAFT	ALLOY STEEL	1
12	107	HANDLE SHAFT	ALLOY STEEL	1
13	106	HANDLE GRIP	RUBBER	1
14	115	CR-BHMS 0.19-32x0.25x0.25-N	ALLOY STEEL	1
15	113	CR-FHMI 0.19-32x0.5x0.5-N	ALLOY STEEL	1

School Name

Kant Twist Clamp

C WPI-1-101-112 A

Exercise 4.9: Arbor Press Project

Use the web to locate a similar assembly and to understand the parts needed in the assembly. View the provided sample parts. Create your own parts and final assembly. View the provided motion file to create the proper movement in SOLIDWORKS.

Create an ANSI Arbor Press assembly document. Insert all needed components and mates to assemble the assembly. Create and insert any additional components if needed.

Create a A-ANSI Landscape - Third Angle Isometric Exploded Drawing document with Explode lines of the assembly using your knowledge of SOLIDWORKS. Insert a BOM with Balloons. Insert all needed General notes and Custom Properties in the Title Block.

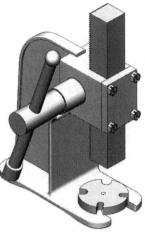

Arbor Press Project
Bench Vice Project
Butterfly Valve Project
Drill Guide Project
Kant Twist Clamp Project
Pipe Vice Project
Pulley Project
Quick Acting Clamp Project
Radial Engine Project
Shaper Tool Head Project
Welder Arm Project

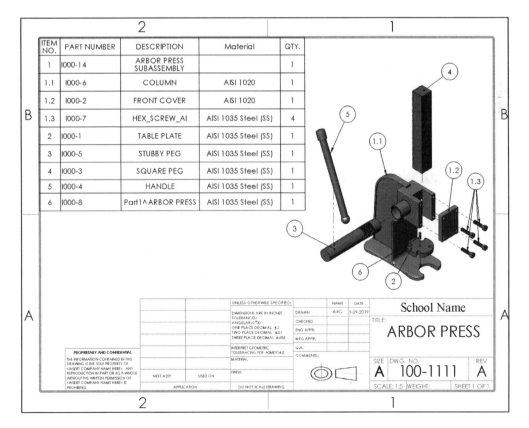

ITEM NO.	PART NUMBER	DESCRIPTION	Material	QTY.
1	I000-14	ARBOR PRESS SUBASSEMBLY		1
1.1	I000-6	COLUMN	AISI 1020	1
1.2	I000-2	FRONT COVER	AISI 1020	1
1.3	I000-7	HEX_SCREW_AI	AISI 1035 Steel (SS)	4
2	I000-1	TABLE PLATE	AISI 1035 Steel (SS)	1
3	I000-5	STUBBY PEG	AISI 1035 Steel (SS)	1
4	I000-3	SQUARE PEG	AISI 1035 Steel (SS)	1
5	I000-4	HANDLE	AISI 1035 Steel (SS)	1
6	I000-8	Part1^ARBOR PRESS	AISI 1035 Steel (SS)	1

UNLESS OTHERWISE SPECIFIED:
DIMENSIONS ARE IN INCHES
TOLERANCES:
ANGULAR: 0°30'
ONE PLACE DECIMAL ±.1
TWO PLACE DECIMAL ±.01
THREE PLACE DECIMAL ±.005

INTERPRET GEOMETRIC TOLERANCING PER: ASMEY14.5

MATERIAL

FINISH

PROPRIETARY AND CONFIDENTIAL
THE INFORMATION CONTAINED IN THIS DRAWING IS THE SOLE PROPERTY OF <INSERT COMPANY NAME HERE>. ANY REPRODUCTION IN PART OR AS A WHOLE WITHOUT THE WRITTEN PERMISSION OF <INSERT COMPANY NAME HERE> IS PROHIBITED.

NAME | DATE
DRAWN | ARG | 7-29-2019
CHECKED
ENG APPR.
MFG APPR.
Q.A.
COMMENTS:

NEXT ASSY | USED ON
APPLICATION | DO NOT SCALE DRAWING

School Name

TITLE:
ARBOR PRESS

SIZE A | DWG. NO. 100-1111 | REV A
SCALE: 1:5 | WEIGHT: | SHEET 1 OF 1

Exercise 4.10: Shaper tool head Project

Use the web to locate a similar assembly and to understand the parts needed in the assembly. View the provided sample parts. Create your own parts and final assembly. View the provided motion file to create the proper movement in SOLIDWORKS.

Create an ANSI Shaper tool head assembly document. Insert all needed components and mates to assemble the assembly. Create and insert any additional components if needed.

Create a C-ANSI Landscape - Third Angle Isometric Exploded Drawing document with Explode lines of the assembly using your knowledge of SOLIDWORKS. Insert a BOM with Balloons. Insert all needed General notes and Custom Properties in the Title Block.

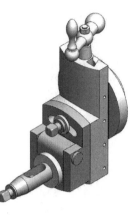

Arbor Press Project
Bench Vice Project
Butterfly Valve Project
Drill Guide Project
Kant Twist Clamp Project
Pipe Vice Project
Pulley Project
Quick Acting Clamp Project
Radial Engine Project
Shaper Tool Head Project
Welder Arm Project

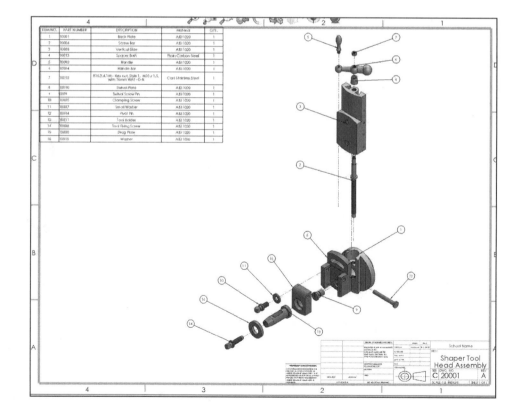

ITEM NO.	PART NUMBER	DESCRIPTION	Material	QTY.
1	10001	Back Plate	AISI 1020	1
2	10006	Screw Bar	AISI 1020	1
3	10003	Vertical Slide	AISI 1020	1
4	10013	Spacer Bush	Plain Carbon Steel	1
5	10002	Handle	AISI 1020	1
6	10004	Handle Bar	AISI 1020	1
7	10012	B16.2.4.1 M - Hex nut, Style 1, M10 x 1.5, with 16mm WAF - D-N	Cast Stainless Steel	1
8	10010	Swivel Plate	AISI 1020	1
9	10009	Swivel Screw Pin	AISI 1020	1
10	10005	Clamping Screw	AISI 1020	1
11	10007	Small Washer	AISI 1020	1
12	10014	Pivot Pin	AISI 1020	1
13	10011	Tool Holder	AISI 1020	1
14	10008	Tool Fixing Screw	AISI 1020	1
15	10000	Drag Plate	AISI 1020	1
16	10015	Washer	AISI 1020	1

Chapter 5

Drawing and Dimensioning Fundamentals

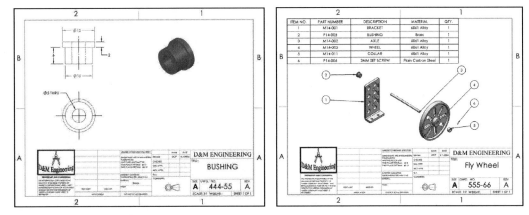

Below are the desired outcomes and usage competencies based on the completion of Chapter 5.

Desired Outcomes:	Usage Competencies:

• An Isometric Exploded Assembly Drawing.
 	
   • Bill of Materials.
 	
   • Balloons.
 	
• A Part Drawing.
 	
   • Front, Top, Right & Isometric view.
 	
   • Dimensions.
 	
   • Annotations.
	• Produce a Sheet Format with Custom Sheet Properties and Title block.
	
	• Create an Exploded Isometric Assembly drawing view with a Bill of Materials and Balloons.
	
	• Create standard Orthographic (Front, Top & Right) views.
	
	• Aptitude to import dimensions from the Model Items tool.
	
	• Knowledge to insert additional dimensions using the Smart Dimension toolbar.
	
	• Ability to insert drawing annotation.
	
	• Modify the Sheet scale.

Notes:

Chapter 5 - Drawing and Dimensioning Fundamentals

Chapter Objective

Create two new drawings: Fly Wheel assembly and Bushing part. The first drawing is an Isometric Exploded view of the Fly Wheel assembly. The assembly drawing displays a Bill of Materials (BOM) at the part level along with Balloons, Magnetic lines and Custom Sheet Properties.

The second drawing is a Bushing part. Insert standard Orthographic (Front, Top & Right) views along with an Isometric view. Insert all needed dimensions, annotations and Sheet Custom Properties. Modify drawing view display styles.

On the completion of this chapter, you will be able to:

- Create a new SOLIDWORKS drawing.

- Set user defined document properties for a drawing.

- Produce a Sheet Format with Custom Sheet Properties.

- Generate an Exploded Isometric Assembly drawing view.

- Create and insert a Bill of Materials (BOM) with Auto Balloons in the Assembly drawing.

- Produce a Part drawing with standard Orthographic (Front, Top & Right) views and an Isometric view.

- Address imported dimensions from the Model Items tool.

- Insert additional dimensions using the Smart Dimension toolbar.

- Modify drawing view display styles.

- Insert drawing view annotations.

- Modify the Sheet scale.

Activity: Start a SOLIDWORKS Session.

Start a SOLIDWORKS session.

1) Double-click the **SOLIDWORKS icon** from the desktop.

2) **Close** the Welcome - SOLIDWORKS dialog box. The Welcome - SOLIDWORKS box provides a convenient way to open recent documents (Parts, Assemblies and Drawings), view recent folders, access SOLIDWORKS resources, and stay updated on SOLIDWORKS news.

The foundation of a new SOLIDWORKS drawing is the Drawing Template. Drawing size, drawing standards, company information, manufacturing, and/or assembly requirements, units and other properties are defined in the Drawing Template.

Activity: Create a New Drawing Document.

Create a New Drawing document.

3) Click **New** 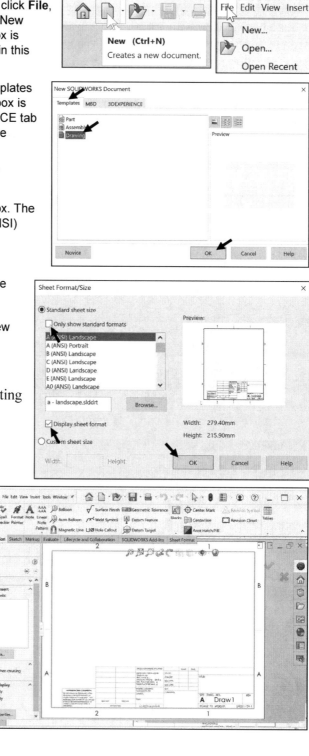 from the Menu bar or click **File**, **New** from the Menu bar menu. The New SOLIDWORKS Document dialog box is displayed. Advanced mode is used in this book.

4) Double-click **Drawing** from the Templates tab. The Sheet Format/Size dialog box is displayed. **Note**: The 3DEXPERIENCE tab is displayed if you are logged into the **3D**EXPERIENCE Platform.

5) If needed, un-check the **Only show standard formats** box.

6) Check the **Display sheet format** box. The default Standard sheet size is A (ANSI) Landscape.

7) Select **A(ANSI) Landscape**.

8) Click **OK** from the Sheet Format/Size dialog box.

9) Click **Cancel** ✖ from the Model View PropertyManager. Draw1 FeatureManager is displayed.

If the Start command when creating new drawing option is checked, the Model View PropertyManager is selected by default.

Draw1 is the default drawing name. Sheet1 is the default first sheet name. The default Drawing

CommandManager tabs and Task Pane tabs will vary depending on system setup and Add-ins.

The Sheet Format is incorporated into the Drawing Template. The Sheet Format contains the border, Title block information, revision block information, company name, and/or company logo information, Custom Properties and SOLIDWORKS Properties. Custom Properties and SOLIDWORKS Properties are shared values between documents.

In the next section, set Sheet Properties in the Assembly drawing document.

Activity: Set Sheet Properties for the Drawing.

Set Sheet Properties for the Drawing.

10) Right-click on **Sheet1** in the Drawing FeatureManager.

11) Click **Properties** 📰 . The Sheet Properties dialog box is displayed.

12) Select **A (ANSI) Landscape** from the Standard sheet size box.

13) Set Sheet Scale: **1:1**.

14) Select **Third angle** for Type of projection.

15) Click **Apply Changes** or **Cancel** from the Sheet Properties dialog box. The Apply Changes command is only available if a change was made. The Sheet Properties are set for the drawing document.

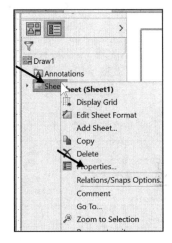

There are two different types of Angle Projection: First and Third Angle Projection.

- First Angle Projection is used primarily in Europe, South American and Asia.

- Third Angle Projection is used in the United States.

Both First Angle and Third Angle projections result in the same six principle views (Front, Top, Right, Bottom, Left and Back); the difference between them is the arrangement of these views.

In the next section, set Document Properties in the drawing.

Activity: Set Document Properties for the Drawing.

Set Document Properties. Select overall drafting standard.

16) Click **Options** ⚙ ▾ from the Menu bar. The System Options General dialog box is displayed.

17) Click the **Document Properties** tab.

18) Select **ANSI** from the Overall drafting standard drop-down menu. Various Detailing options are available depending on the selected standard.

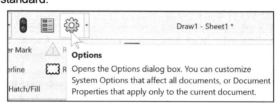

The Overall drafting standard determines the display of dimension text, arrows, symbols and spacing.

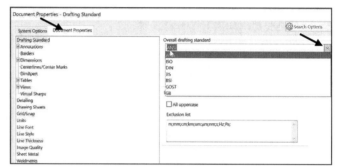

Units are the measurement of physical quantities. Millimeter dimensioning and decimal inch dimensioning are the two most common unit types specified for engineering parts and drawings.

Set Document Properties. Select units and precision.

19) Click the **Units** folder.

20) Click **MMGS** (millimeter, gram, second) for Unit system.

21) Select **.12** (two decimal places) for Length basic units.

22) Click **OK** from the Document Properties - Units dialog box. The Part FeatureManager is displayed.

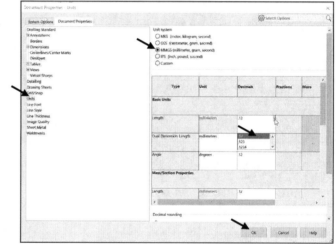

Document Properties control the display of dimensions, annotations and symbols in the drawing. Sheet Properties control Sheet Size, Sheet Scale and Type of Projection.

Title Block

The Title block contains text fields linked to System Properties and Custom Properties. System Properties are determined from the SOLIDWORKS documents. Custom Property values are assigned to named variables. Save time. Utilize System Properties and define Custom Properties in your Sheet Formats.

System Properties and Custom Properties for Title Block:			
System Properties Linked to fields in default Sheet Formats:	**Custom Properties of drawings linked to fields in default Sheet Formats:**		**Custom Properties of parts and assemblies linked to fields in default Sheet Formats:**
SW-File Name (in DWG. NO. field):	CompanyName:	EngineeringApproval:	Description (in TITLE field):
SW-Sheet Scale:	CheckedBy:	EngAppDate:	Weight:
SW-Current Sheet:	CheckedDate:	ManufacturingApproval:	Material:
SW-Total Sheets:	DrawnBy:	MfgAppDate:	Finish:
	DrawnDate:	QAApproval:	Revision:
	EngineeringApproval:	QAAppDate:	

The drawing document contains two Sheet modes:

- Edit Sheet Format

- Edit Sheet

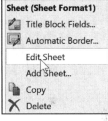

Insert views and dimensions in the Edit Sheet mode. Modify the Sheet Format text, lines and Title block information in the Edit Sheet Format mode.

The CompanyName Custom Property is located in the Title block above the TITLE box. There is no value defined for CompanyName. A small text box indicates an empty field. Define a value for the Custom Property CompanyName. Example: D&M EDUCATION LLC. The Tolerance block is located in the Title block.

A part cannot be inserted into a drawing when the Edit Sheet Format mode is selected.

The Tolerance block provides information to the manufacturer on the minimum and maximum variation for each dimension on the drawing.

If a specific tolerance or note is provided on the drawing, the specific tolerance or note will override the information in the Tolerance block. General tolerance values are based on the design requirements and the manufacturing process. Modify the Tolerance block in the Sheet Format for ASME Y14.5 machined parts. Delete unnecessary text. The FRACTIONAL text refers to inches. The BEND text refers to sheet metal parts.

Fly Wheel Assembly Drawing

A drawing contains part views, geometric dimensioning and tolerances, notes, and other related design information. When a part is modified, the drawing automatically updates. When a driving dimension in the drawing is modified, the part is automatically updated.

Drawings consist of one or more views produced from the part or assembly. Create the Fly Wheel Assembly drawing from the Fly Wheel assembly.

Utilize the Task Pane to insert an Isometric Exploded drawing view. Create a Bill of Materials at the part level. Unlike a part drawing, no dimensions are required at this time. Utilize the SOLIDWORKS 2024\Drawing folder for models in the section.

Activity: Create a Fly Wheel Assembly Drawing.

Insert the Isometric Exploded Drawing view of the Fly Wheel Assembly.

23) Click the **View Palette** ⊞ icon from the SOLIDWORKS Resource Pane.

24) Click the **Browse** button. The Windows Open dialog box is displayed.

25) Double-click the **Fly Wheel** assembly from the SOLIDWORKS 2024\Drawing folder.

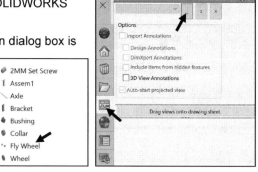

Insert the Isometric Exploded view.

26) **Drag and drop** the *Isometric Exploded view icon into Sheet1. The view is displayed. **Note**: The Isometric Exploded view was created in Chapter 4.

27) Click the **Green Check mark** ✔ from the Drawing PropertyManager.

Set Sheet Properties for the Drawing.

28) **Right-click** Sheet1 in the Fly Wheel FeatureManager as illustrated.

29) Click **Properties** ▤. The Sheet Properties dialog box is displayed.

30) Enter Scale: **1:2**.

31) Click **Apply Changes** from the Sheet Properties dialog box.

32) **View** the results in the Graphics window.

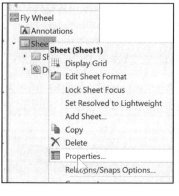

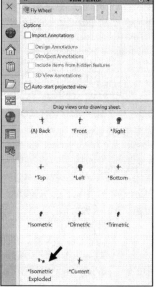

Modify the Display Style.

33) Click **inside** the Exploded Isometric drawing view. The Drawing View1 PropertyManager is displayed.

34) Click **Shaded with Edges** for Display Style.

35) Click the **Green Check mark** ✔ from the Drawing View1 PropertyManager.

Save the Fly Wheel assembly drawing.

36) Click **Save As** 📄.

37) **Select** the save in folder SOLIDWORKS 2024\Drawing folder. The assembly, parts and drawing are saved in the same folder.

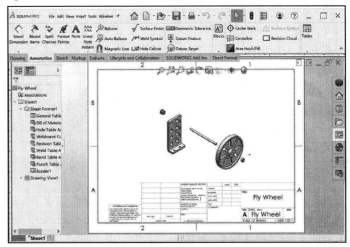

38) Click **Save**.

Balloons

You do not have to insert a Bill of Materials in order to add balloons. Use the Auto Balloon tool to automatically generate balloons in a drawing view.

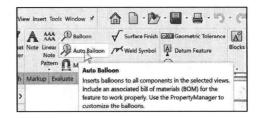

In the Graphics area, click an assembly component and then click to place the balloon.

Insert multiple balloons without closing the PropertyManager.

When the drawing contains a Bill of Materials, use Item Numbers to set the balloon sequence.

In the next section, insert Auto Balloons using Magnetic lines in the assembly.

Activity: Insert Auto Balloons into the Assembly Drawing.

Insert Auto Balloons with Magnetic lines.

39) Click inside the **Exploded Isometric** drawing view. The Drawing View 1 PropertyManager is displayed.

40) Click the **Annotation** tab from the CommandManager.

41) Click the **Auto Balloon** icon from the Annotation tab in the CommandManager.

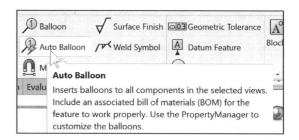

Auto Balloon
Inserts balloons to all components in the selected views. Include an associated bill of materials (BOM) for the feature to work properly. Use the PropertyManager to customize the balloons.

42) **View** the balloons in the Graphics window. **Note**: Your balloon locations may vary.

43) Click **OK** from the Auto Balloon PropertyManager.

44) **View** the results.

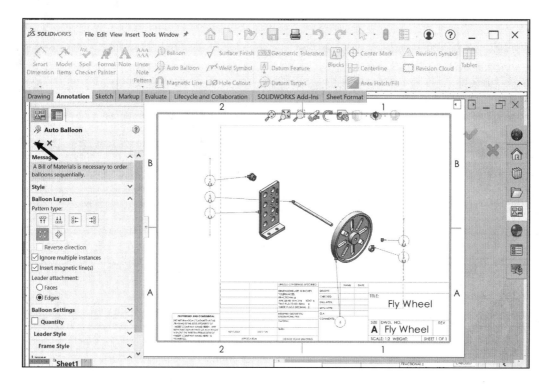

45) **Click** and **drag** the Balloons as illustrated.

46) Click **OK** ✔ from the Balloon PropertyManager.

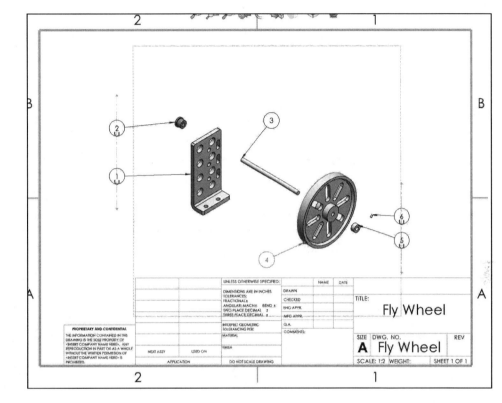

💡 Magnetic lines are a convenient way to align balloons along a line at any angle. You attach balloons to magnetic lines, choose to space the balloons equally or not and move the lines freely at any angle in the drawing.

Bill of Materials

SOLIDWORKS automatically populates a Bill of Materials (BOM) with item numbers, quantities, part numbers and custom properties in assembly drawings if they are inserted in the proper document. You can anchor, move, edit and split a BOM.

When you insert balloons into a drawing, the item numbers and quantities in the balloons correspond to the numbers in the Bill of Materials.

If an assembly has more than one configuration, you can list quantities of components for all configurations or selected configurations.

You can create BOMs in assembly files and multi-body part files. You can insert a BOM saved with an assembly into a referenced drawing. You do not need to create a drawing first.

In the next section, insert a Bill of Materials (BOM).

Activity: Insert a Bill of Materials.

Insert a Bill of Materials at the part level.

47) Click inside the **Exploded Isometric view** in the Graphics window.

48) Click the **Annotation** tab from the CommandManager.

49) Click the **Tables** drop-down menu.

50) Click the **Bill of Materials** icon. The Bill of Materials PropertyManager is displayed.

51) Click the **Open table template for Bill of Materials** icon.

52) Double-click **bom-material.sldbomtbt**.

53) Check the **Parts only** BOM Type.

54) Click **OK** from the Bill of Materials PropertyManager.

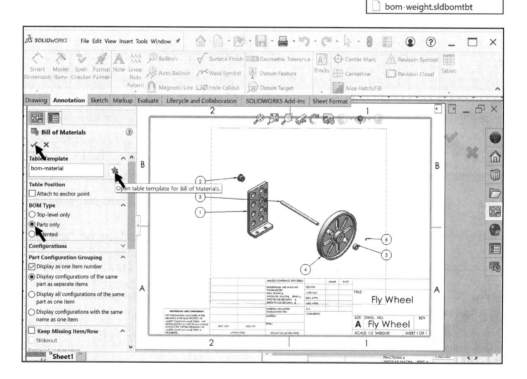

55)   Click a **position** in the upper left corner to place the Bill of Material Table.

56)   **View** the results. Move the Isometric Exploded view if needed. Size the cells in the Bill of Material table if needed.

Save the Fly Wheel assembly drawing.

57)   Click **Save** 💾.

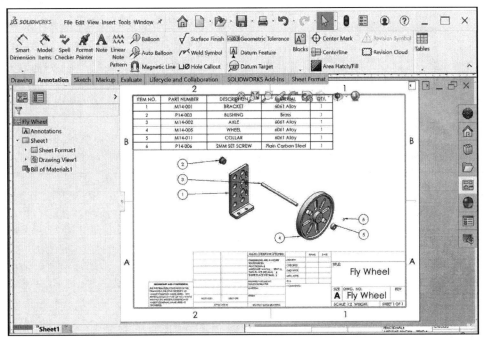

CommandManager tabs will vary depending on system setup and Add-ins.

In the next section address Custom Properties of the parts in the Bill of Materials.

Later, address Custom Properties of the Drawing and Title block.

For a Part Document

For a Drawing Document

Activity: Address Custom Properties of the Parts in the Bill of Materials.

Open the Axle part from the drawing.

58) Right-click the **Axle** in the drawing view.

59) Click **Open Part**. The Axle part is displayed. The Axle FeatureManager is displayed. Material is 6061 Alloy. For the Material to be displayed in a drawing or Bill of Materials, you need to set the Custom Properties for each part in an assembly along with other required information for the drawing. Note: In this case, it's set for you.

View Custom Properties for the Axle part.

60) Click **File**, **Properties** from the Menu bar. The Summary Information dialog box is displayed.

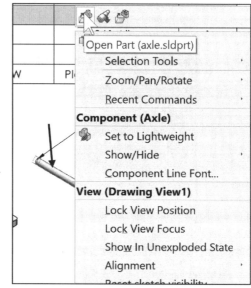

61) Click the **Configuration Specific** tab. View the entered information: *Description* and *Material*. You would need to enter this information for all new created parts to populate the BOM and Title box in a drawing. The entered Description and Material information is related to the drawing document Hyperlinks.

Note: <u>Save to This PC</u>: Saves the document to your local hard drive in the folder which was last opened from. <u>Save to 3DEXPERIENCE</u>: Opens the Save as New dialog box to save the file to a Collaborative space on the **3D**EXPERIENCE platform. The Save to 3DEXPERIENCE option is displayed if you purchased 3DEXPERIENCE education roles and are logged into the **3D**EXPERIENCE platform. <u>Save As</u>: Saves the document to a new file name that becomes the active document without saving the original document. <u>Save All</u>: Saves all the open files that have been modified since they were last saved.

3DEXPERIENCE users can create templates for the **3D**EXPERIENCE platform directly from SOLIDWORKS. Use the 3DEXPERIENCE tab in the New SOLIDWORKS Document dialog box.

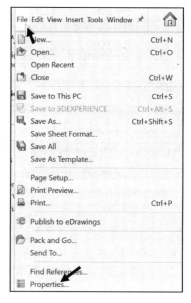

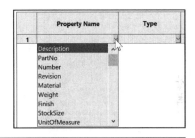

Summary Information

Summary | Custom | **Configuration Specific**

Apply to: BOM quantity:

[Delete]   Default ⌄   - None - ⌄   [Edit List]

	Property Name	Type	Value / Text Expression	Evaluated Value
1	Description	Text	AXLE	AXLE
2	Material	Text	"SW-Material@@Default@Axle.SLDPRT"	6061 Alloy
3	<Type a new property>			

As an exercise, delete Row1 & 2. Re-enter Description and Material information.

62) Click in **Row 2**. Press the **Delete** button.

63) Click in **Row1**. Press the **Delete** button.

64) Click the **drop-down arrow** under Property Name as illustrated.

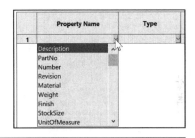

	Property Name	Type
1	Description	

Description
PartNo
Number
Revision
Material
Weight
Finish
StockSize
UnitOfMeasure

65) Select **Description**.

66) Click inside the **Value / Text Expression** box.

67) Enter **AXLE**.

68) Click the **drop-down arrow** under Property Name in Row 2.

Summary Information

Summary | Custom | Configuration Specific

Apply to:

[Delete]   Default ⌄

	Property Name	Type	Value / Text Expression
1	Description	Text	AXLE
2	<Type a new property>		

69) Select **Material**.

70) Click the **drop-down arrow** under Value / Text Expression.

71) Select **Material**. View the results. The Material drawing link is displayed. 6061 Alloy is displayed in the Evaluated Value box.

	Property Name	Type	Value / Text Expression
1	Description	Text	AXLE
2	Material	Text	
3	<Type a new property>		Material

Material
Mass
Density
Volume

72) Click **OK** from the Summary Information dialog box.

73) **Close** the AXLE part.

74) **Return** to the drawing. Address Custom Properties for the drawing.

Summary Information

Summary | Custom | Configuration Specific

Apply to: BOM quantity:

[Delete]   Default ⌄   - None - ⌄   [Edit List]

	Property Name	Type	Value / Text Expression	Evaluated Value
1	Description	Text	AXLE	AXLE
2	Material	Text	"SW-Material@@Default@Axle.SLDPRT"	6061 Alloy
3	<Type a new property>			

Activity: Address Custom Properties - Modify the Title block.

Define Custom Properties in the drawing document.

75) Click **File**, **Properties** from the Menu bar. The Summary Information dialog box is displayed.

76) Click the **Custom** tab.

77) **Select and enter** the illustrated items. Note: Enter the present DrawnDate and your DrawnBy initials. Enter your school's name under CompanyName.

78) Click **OK** from the Summary Information dialog box. View the results.

Activate the Edit Sheet Format mode.

79) **Right-click** in Sheet1. Do not click inside the Isometric drawing view.

80) Click **Edit Sheet Format**. The Title block lines turn blue.

81) **Zoom in** on the Title block.

Modify the Tolerance Note.

82) Double-click the text **INTERPRET GEOMETRIC TOLERANCING PER:**.

83) Enter **ASME Y14.5**.

84) Click **OK** ✔ from the Note PropertyManager.

85) Double-click inside the **Tolerance block** text. The Formatting dialog box and the Note PropertyManager is displayed.

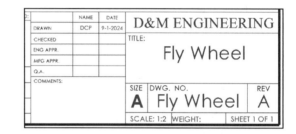

86) Delete the text **INCHES**.

87) Enter **MILLIMETERS**.

88) Delete the line **FRACTIONAL ±** as illustrated.

89) Delete the text **BEND ±** as illustrated.

90) Click a position at the end of the **ANGULAR: MACH ±** line.

91) Enter **0**.

92) Click the **Add Symbol** button from the Text Format box. The Symbols dialog box is displayed.

93) Click **Degree** from the drop-down menu.

94) Enter **30′** for minutes of a degree.

95) Delete the **TWO** and **THREE PLACE DECIMAL** lines.

96) Enter **ONE PLACE DECIMAL ±0.5**.

97) Enter **TWO PLACE DECIMAL ±0.15**.

98) Click **OK** ✔ from the Note PropertyManager.

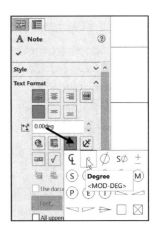

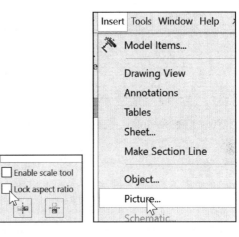

Insert a Company Logo and the Third Angle Projection icon.

99) Click **Insert**, **Picture** from the Menu bar.

100) Double-click the **LOGO.jpg** from the Drawing folder. Note, if you have your own logo, use it for the drawing.

101) The Sketch Picture PropertyManager is displayed. Un-check the **Lock aspect ratio** box and the **Enable scale tool** box.

102) Drag the **picture handles** to size the picture to the left of the Title block as illustrated.

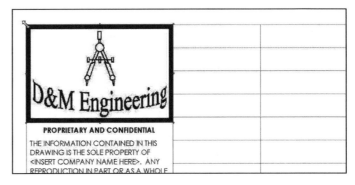

103) Click **OK** ✔ from the Sketch Picture PropertyManager.

104) Click **Insert**, **Picture** from the Menu bar. The Open dialog box is displayed.

105) Double-click the **Third Angle Projection.jpg**. The Sketch Picture PropertyManager is displayed.

106) Drag the **picture handles** to size the icon.

107) Click **OK** ✔ from the Sketch Picture PropertyManager.

Return to the Edit Sheet mode.

108) **Right-click** in the Graphics window.

109) Click **Edit Sheet**. View the results.

Save the Fly Wheel assembly drawing.

110) Click **Save** 💾. As an exercise, change the DWG. NO to 555-66 as shown below in the Edit Sheet Format Mode. Update the DrawnBy and DrawnDate in the Title box. Apply Custom Properties and Link Properties to the drawing.

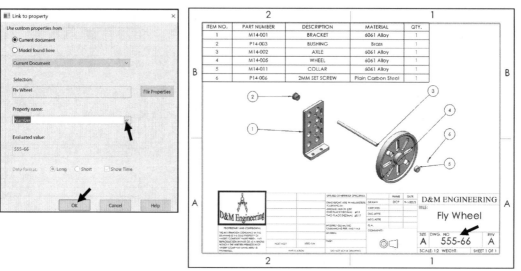

Bushing Part Drawing

A drawing contains part views, geometric dimensioning and tolerances, notes, and other related design information. When a part is modified, the drawing automatically updates. When a driving dimension in the drawing is modified, the part is automatically updated.

Create a Bushing part drawing with inserted dimensions and annotations. Modify the Display mode to show Hidden Lines.

Drawings consist of one or more views produced from the part or assembly. Insert four drawing views: Front, Top, Right and Isometric. Hide any unnecessary drawing views. The Front view should be the most descriptive view. It should be positioned in a natural orientation based on its function.

Open and start the Bushing Drawing with an empty Bushing Drawing document located in the SOLIDWORKS 2024\Drawing folder.

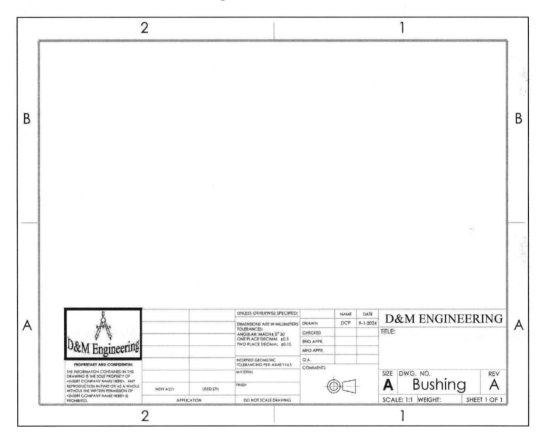

Activity: Open the Empty Bushing Part Drawing.

Open the Empty Bushing Part Drawing.

111) Click **File**, **Open** from the Main menu.

112) Browse to the **SOLIDWORKS 2024\Drawing** folder.

113) Double-click the **Bushing** part drawing. The empty Bushing part drawing is displayed. Third Angle Projection is selected for Projection type. ANSI is the drafting Standard. MMGS are the units.

There are two different types of Angle Projection: First and Third Angle Projection.

- First Angle Projection is used in Europe, South American and Asia.

- Third Angle Projection is used in the United States.

Both First Angle and Third Angle

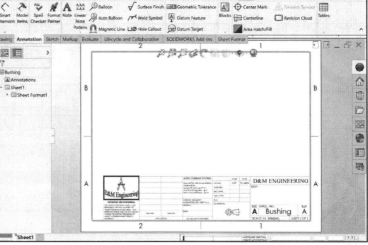

projections result in the same six principle views (Front, Top, Right, Bottom, Left and Back); the difference between them is the arrangement of these views.

Insert four drawing views: Front, Top, Right and Isometric. Utilize the View Palette in the Task Pane.

All drawing views are projected from the Front view. You can place any view in the Front view location.

The Front view should be the most natural and descriptive view of the model.

Insert four drawing views.

114) Click the **View Palette** ⊞ icon from the SOLIDWORKS Resources Pane.

115) Click the **Browse** button.

116) Browse to the **SOLIDWORKS 2024\Drawing** folder.

117) Double-click the **Bushing** part.

118) **Drag** and **drop** the (A) Right view into the Front view location as illustrated.

119) Click directly **upwards** to create the Top view.

120) Click directly **downward** and to the **right** to create the Right view.

121) Click a location **approximately 45 degrees** to the upper right of the Front view. This is the Isometric view.

122) Click **OK** ✔ from the PropertyManager. View the four views.

123) Click and drag the **drawing views** to space them for dimensions and annotations. The Top and Right view are directly projected from the Front view.

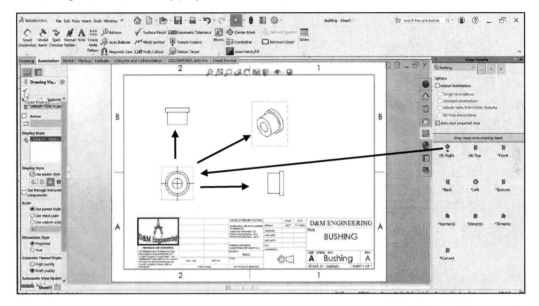

Save the Bushing part drawing in the SOLIDWORKS 2024\Drawing folder.

124) Click **Save As** .

125) Select the **SOLIDWORKS 2024\Drawing** folder.

126) Click **Save**. Click **Yes**. The part document and drawing document are saved in the same folder.

Drawing Dimensions

Dimension the part drawing views. Utilize the Model items tool to automatically import model dimensions. Utilize the Smart Dimension tool to insert any additional needed dimensions.

In SOLIDWORKS, inserted dimensions in the drawing are displayed in gray. Imported dimensions from the part are displayed in black.

Import Model dimensions using the Model Items tool.

127) Click the **Annotation** tab.

128) Click the **Model Items** tool. The Model Items PropertyManager is displayed. View your options and tools.

129) Select **Entire model** from the Source drop-down menu.

130) Click the **Hole Wizard Locations** icon.

131) Click the **Hole callout** icon. This option inserts hole callout annotations to hole wizard features.

132) Click **OK** from the Model Items PropertyManager. Imported part dimensions are displayed in the drawing.

Dimensions are imported into the drawing. The dimensions may not be in the correct location with respect to the feature or size of the part per the ANSI Y-14.5 standard. Move them later in the chapter and address extension line gaps (5mm) and annotations (Centerline).

Move parent (Front) and child (Top and Right) views independently by dragging their view boundary. Hold the Shift key down and select multiple views to move as a group.

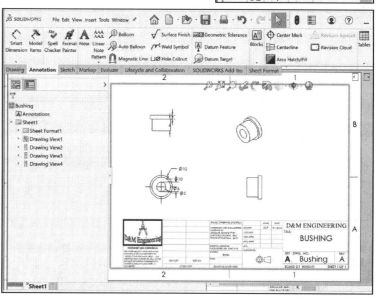

Third Angle Projection type symbol is illustrated.

The dimensions and text in the next section have been enlarged for visibility. Drawing dimension location is dependent on *Feature dimension creation* and *Selected drawing views.*

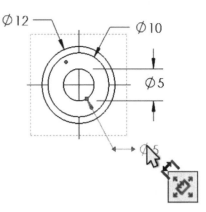

Move dimensions within the same view. Use the mouse pointer to drag dimensions and leader lines to a new location. Leader lines reference the size of the profile. A gap must exist between the profile lines and the leader lines. Shorten the leader lines to maintain a drawing standard. Use the blue Arrow buttons to flip the dimension arrows.

Notes provide relative part or assembly information. Example: Material type, material finish, special manufacturing procedure or considerations, preferred supplier, etc.

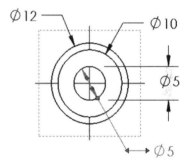

Below are a few helpful guidelines to create general drawing notes:

- Use Upper Case letters.

- Use left text justification.

- Font size should be the same as the dimension text.

Hide superfluous feature dimensions. Do not delete feature dimensions. Recall hidden dimensions with the View, Show Annotations command. Move redundant, dependent views outside the sheet boundary or hide redundant views.

Full cylinders (holes and bosses) must always be measured by their diameter. The diameter symbol must precede the numerical value to indicate that the dimension shows the diameter of a circle or cylinder. The symbol used is the Greek letter phi (Ø).

If a hole goes completely through the feature and it is not clearly shown on the drawing, the abbreviation "**THRU**" or "**THRU ALL**" in all upper case follows the dimension.

You should not dimension a hole in a drawing by its radius.

Move and Hide drawing dimensions in the Front view.
133) Zoom-in on the Front view.

134) Right-click the bottom **Ø5** in the Front view.

135) Click **Hide**. Note: Expand the drop-down menu to view the Hide command.

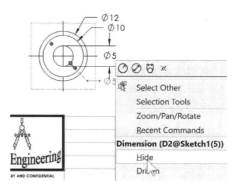

136) Right-click **Ø12** in the Front view.

137) Click **Hide**.

138) Right-click **Ø10** in the Front view.

139) Click **Hide**.

140) Right-click the second **Ø5** dimension.

141) Click **Hide**. No dimensions are displayed in the Front view.

142) Click the **Annotations** tab. Click the **Smart Dimension** tool.

143) Insert a **Ø5** dimension in the Front view with a bent leader line as illustrated.

144) Click a **position after** <MOD-DIM><DIM> in the Dimension Text box.

145) Press the **space** key.

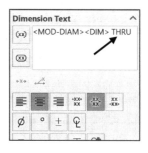

146) Enter **THRU**. Click **OK** ✔ from the Dimension PropertyManager. View the results in Drawing view1.

Display Hidden Lines Visible.
147) Click **inside Drawing View2** as illustrated. The Drawing View2 PropertyManager is displayed.

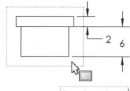

148) **Move** the dimensions as illustrated.

149) Click **Hidden Lines Visible** ⬡ from the Display Style box.

150) Click **OK** ✔ from the Drawing View2 PropertyManager. View the results.

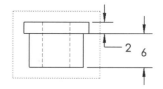

💡 General practice is to stagger the dimension text on parallel dimensions (small to large).

💡 Arrowheads are used to terminate dimension lines.

Insert needed feature and location dimensions in the Top view.

151) Zoom-in on the Top view.

152) Click the **6mm** dimension.

153) Move the 6mm dimension to the left of the drawing view as illustrated.

154) Click the **control points** on the extension lines to create 5mm gaps as illustrated.

💡 All drawing views with dimensions require gaps (5mm) between the visible feature line and extension lines.

Insert a diameter dimension in the Top View.

155) Click **Smart Dimension** ◇ from the CommandManager.

156) Click the **top horizontal line** as illustrated.

157) Click **position** above the model. View the results.

Insert a second diameter dimension in the Top View.

158) Click the **bottom horizontal line**.

159) Click a position **below** the model. View the results.

160) Click **OK** ✓ from the Dimension PropertyManager

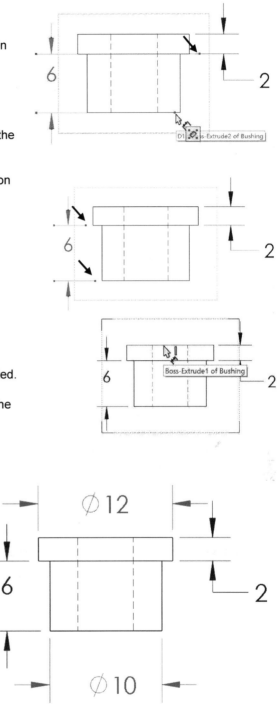

Centerlines or Center marks should be used on all circles, holes and slots.

Insert a Centerline in the Top View.

161) Click **inside Drawing View2**. The Drawing View2 PropertyManager is displayed.

162) Click **Centerline** ⊞ from the Annotation CommandManager. The Centerline PropertyManager is displayed.

163) Click the **Select View** box.

164) Click **OK** ✔ from the Centerline PropertyManager. View the results. Move any dimensions if needed.

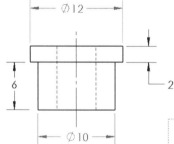

Display Shaded With Edges (Drawing View4).

165) Click **inside Drawing View4** as illustrated. The Drawing View4 PropertyManager is displayed.

166) Click **Shaded with Edges** 🔲 from the Display Style box.

167) Click **OK** ✔ from the Drawing View4 PropertyManager. View the results.

Save the Bushing part drawing.

168) Click **Save** 💾.

169) Click **Yes**.

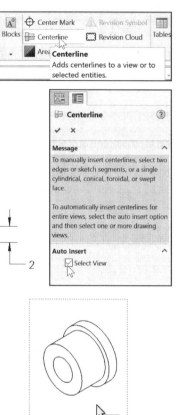

Hide the Right View.

170) Right-click in **Drawing View3**.

171) **Expand** the drop-down menu.

172) Click **Hide**. Hide unneeded drawing views. The Right View is not needed.

173) Click **OK** ✔ from the Drawing View3 PropertyManager. View the results.

Save the Bushing part drawing.

174) Click **Save** 💾.

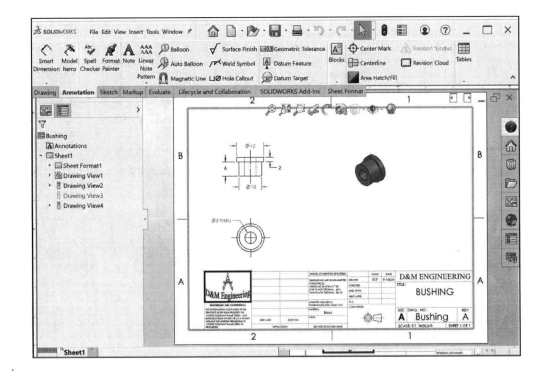

💡 Drawing views and notes outside the sheet boundary do not print.

Modify the Sheet Scale.

175) **Right-click** Sheet1 in the Bushing Drawing FeatureManager.

176) Right-click **Properties**. The Sheet Properties dialog box is displayed.

177) Enter Scale **3:1**.

178) Click **Apply Changes**.

Save the Bushing part drawing.

179) Click **Save** . As an exercise, change the DWG. NO to 444-55 as shown below in the Edit Sheet Format mode. Update the DrawnBy and DrawnDate in the Title box. Apply Custom Properties and Link Properties to the drawing.

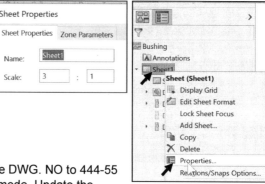

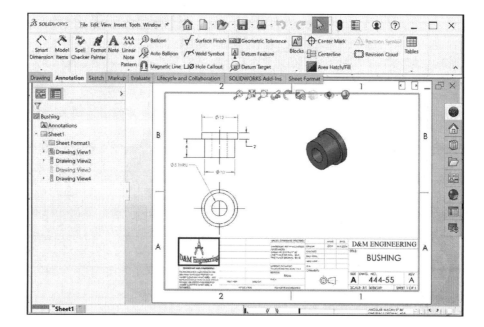

Summary

You learned about the Drawing and Dimension Fundamentals and created two new drawings with user defined document properties:

- Fly Wheel Assembly.

- Bushing.

You created the Fly Wheel Assembly drawing with an Exploded Isometric view.

You utilized a Bill of Materials, Magnetic lines, Balloons and learned about Custom Properties and the Title Block.

You created the Bushing Part drawing utilizing Third Angle Projection with standard Orthographic views: Front, Top, Right and Isometric.

You addressed imported dimensions from the Model Items tool and then inserted additional dimensions using the Smart Dimension tool along with all needed annotations.

The Sheet scale was modified.

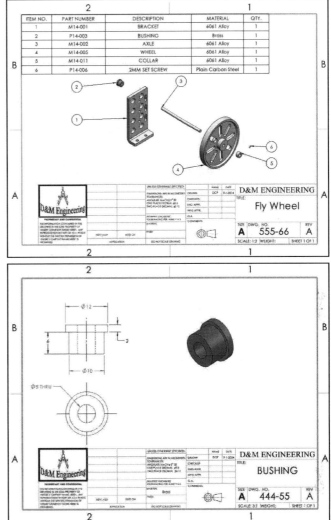

A multi-view drawing should have the minimum number of views necessary to describe an object completely. The most descriptive views are those that reveal the most information about the features, with the fewest features hidden from view.

Questions

1. Describe a Bill of Materials (BOM) in a SOLIDWORKS drawing.

2. Name the two major sheet design modes in a SOLIDWORKS drawing.

3. True or False. Units, Dimensioning Standards, Arrow size, Font size are modified in the Options, Document Properties section.

4. Describe the procedure to save a Sheet Format.

5. Name three components that are commonly found in a title block.

6. In SOLIDWORKS, a drawing file name ends with a _____ suffix.

7. In SOLIDWORKS, a part file name ends with a _____ suffix.

8. True or False. In SOLIDWORKS, if a part is modified, the associated drawing is updated.

9. True or False. In SOLIDWORKS, when a dimension in the drawing is modified, the associated part is updated.

10. Name two guidelines to create General Notes (ASME Y14.5) in a SOLIDWORKS drawing.

11. True or False. Most engineering drawings (ASME Y14.5) use the following font: Times New Roman - all small letters.

12. Describe a Leader line in a SOLIDWORKS drawing.

13. Describe a Detail view in SOLIDWORKS.

14. Describe a Section view in SOLIDWORKS.

15. True or False. In SOLIDWORKS, inserted dimensions in the drawing are displayed in gray. Imported dimensions from the part are displayed in black.

16. True or False. The default SOLIDWORKS Drawing Templates contain predefined Title block Notes linked to Custom Properties and SOLIDWORKS Properties.

17. True or False. In SOLIDWORKS, Draw1 is the default drawing name. Sheet1 is the default first sheet name.

Exercises

Exercise 5.1: FLATBAR - 3 HOLE Drawing

Create the A (ANSI) Landscape - IPS - Third Angle 3HOLES drawing as illustrated below. Do not display Tangent Edges. Do not dimension to Hidden Lines.

- First create the part from the drawing, then create the drawing. Use the default A (ANSI) Landscape Sheet Format/Size.

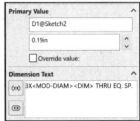

- Insert a Front, Top and a Shaded Isometric view as illustrated. Insert dimensions. Address extension line gaps. Address proper display modes.

- Add a Smart (Linked) Parametric note for MATERIAL THICKNESS in the drawing as illustrated. Hide the dimension in the Top view. Insert Centerlines.

- Modify the Hole dimension text to include 3X THRU EQ. SP. and 2X as illustrated.

- Insert Company and Third Angle projection icons. The icons are available in the homework folder.

- Insert Custom Properties: Description, Material, DrawnBy, CompanyName, Number, Revision and DrawnDate.

- Material is 1060 Alloy.

	Property Name	Type	Value / Text Expression	Evaluated Value
1	Number	Text	556-099	556-099
2	DrawnBy	Text	DCP	DCP
3	DrawnDate	Text	9-1-2024	9-1-2024
4	CompanyName	Text	D&M ENGINEERING	D&M ENGINEERING
5	Description	Text	3HOLES	3HOLES
6	SWFormatSize	Text	8.5in*11in	8.5in*11in

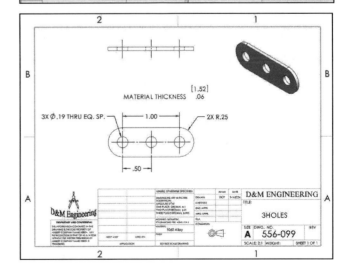

Exercise 5.2: CYLINDER Drawing

Create the A (ANSI) Landscape - IPS - Third Angle
CYLINDER drawing as illustrated below. Do not display
Tangent Edges. Do not dimension to Hidden lines.

- First create the part from the drawing - then create the
  drawing. Use the default A (ANSI) Landscape Sheet
  Format/Size.

- Insert the Front, Right and Isometric
  view as illustrated. Insert dimensions.
  Think about the proper view for your
  dimensions. Address extension line
  gaps. Insert proper display modes.

- Insert Company and Third Angle
  projection icons. The icons are
  available in the homework
  folder.

- Insert Centerlines, Center
  Marks and Annotations.

- Insert Custom Properties:
  Material, Description,
  DrawnBy, CompanyName,
  Number, Revision, and
  DrawnDate. Material is AISI
  1020.

Sheet Format/Size

- ● Standard sheet size
- ☐ Only show standard formats

| A (ANSI) Landscape |
| A (ANSI) Portrait |
| B (ANSI) Landscape |
| C (ANSI) Landscape |
| D (ANSI) Landscape |
| E (ANSI) Landscape |
| A0 (ANSI) Landscape |

a - landscape.slddrt   Browse...

	Property Name	Type	Value / Text Expression	Evaluated Value
1	Material	Text	"SW-Material@CylinderPart1.SLDPRT"	AISI 1020
2	Description	Text	CYLINDER	CYLINDER

	Property Name	Type	Value / Text Expression	Evaluated Value
1	DrawnDate	Text	9-1-2024	9-1-2024
2	DrawnBy	Text	DCP	DCP
3	CompanyName	Text	D&M ENGINEERING	D&M ENGINEERING
4	Revision	Text	A	A
5	Number	Text	667-888	667-888
6	SWFormatSize	Text	8.5in*11in	8.5in*11in
7				

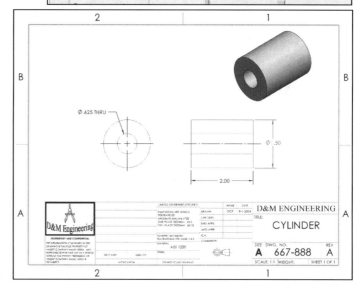

Exercise 5.3: PRESSURE PLATE Drawing

Create the A (ANSI) Landscape - IPS - Third Angle
PRESSURE PLATE drawing. Do not display Tangent edges.
Do not dimension to hidden lines.

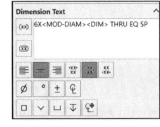

- First create the part from the drawing, then create the
  drawing. Use the default A (ANSI) Landscape Sheet
  Format/Size.

- Insert the Front and Right view as
  illustrated. Insert dimensions.
  Address all extension line gaps.
  Think about the proper view for
  your dimensions. Address proper
  display modes.

- Insert Company and Third
  Angle projection icons. The
  icons are available in the
  homework folder.

- Insert Centerlines, Center
  Marks and Annotations.

- Insert Custom Properties:
  Description, Material,
  DrawnBy,
  CompanyName,
  Number,
  Revision, and
  DrawnDate.

- Material is 1060
  Alloy.

Property Name	Type	Value / Text Expression	Evaluated Value
DrawnDate	Text	9-1-2024	9-1-2024
CompanyName	Text	D&M ENGINEERING	D&M ENGINEERING
Number	Text	55-568	55-568
Revision	Text	A	A
Description	Text	PRESSURE PLATE	PRESSURE PLATE
SWFormatSize	Text	8.5in*11in	8.5in*11in

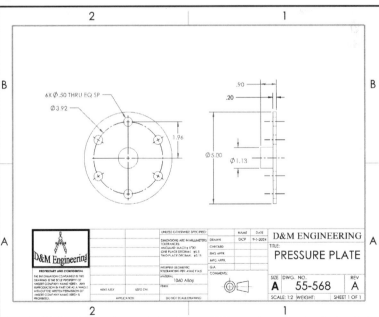

Exercise 5.4: LINKS Assembly Drawing

- Create the LINK assembly. Utilize three different FLATBAR configurations and a SHAFT-COLLAR. The parts are in the Chapter 5 homework folder.

- Create the LINK assembly drawing as illustrated. Use the default A (ANSI) Landscape Sheet Format/Size.

- Insert Company and Third Angle projection icons. The icons are available in the homework folder.

- Remove all Tangent Edges.

- Insert Custom Properties: Description, DrawnBy, CompanyName, Number, Revision and DrawnDate.

	Property Name	Type	Value / Text Expression	Evaluated Value
2	DrawnDate	Text	9-1-2024	9-1-2024
3	CompanyName	Text	D&M ENGINEERING	D&M ENGINEERING
4	Revision	Text	A	A
5	Description	Text	LINK-2	LINK-2
6	SWFormatSize	Text	215.9mm*279.4mm	215.9mm*279.4mm
7	Number	Text	9998-099	9998-099

- Insert a Bill of Materials as illustrated with Balloons.

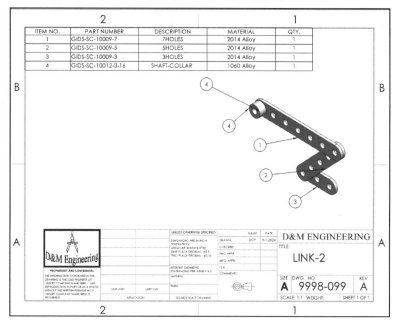

Exercise 5.5: PLATE-1 Drawing

Create the A (ANSI) Landscape - MMGS - Third Angle PLATE-1 drawing as illustrated below. Do not display Tangent edges. Do not dimension to Hidden lines.

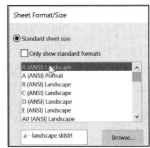

- First create the part from the drawing, then create the drawing. Use the default A (ANSI) Landscape Sheet Format/Size.

- Insert the Front and Right view as illustrated. Insert dimensions. Address extension line gaps. Think about the **proper view** for your dimensions. Address display modes.

- Insert Company and Third Angle projection icons. The icons are available in the homework folder.

- Insert Centerlines, Center Marks and Annotations.

- Insert Custom Properties: Description, Material, DrawnBy, CompanyName, Number, Revision and DrawnDate.

- Material is 1060 Alloy.

	Property Name	Type	Value / Text Expression	Evaluated Value
2	DrawnBy	Text	DCP	DCP
3	DrawnDate	Text	9-1-2024	9-1-2024
4	CompanyName	Text	D&M ENGINEERING	D&M ENGINEERING
5	Revision	Text	A	A
6	Number	Text	5444-999	5444-999
7	SWFormatSize	Text	215.9mm*279.4mm	215.9mm*279.4mm
8				

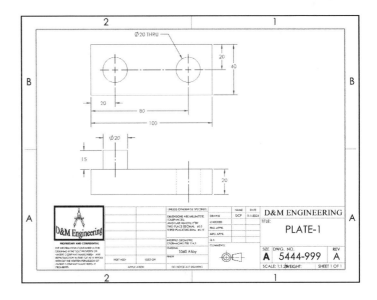

Exercise 5.6: **PLATE Drawing**

Create the A (ANSI) Landscape - IPS - Third Angle FLAT-PLATE drawing. Do not display Tangent edges. Phantom lines are fine.

- First create the part from the drawing, then create the drawing. Use the default A (ANSI) Landscape Sheet Format/Size.

- Insert the Front, Top, Right and Isometric views as illustrated. Insert dimensions. Think about the proper view for your dimensions.

- Address proper display modes.

- Insert Company and Third Angle projection icons. The icons are available in the homework folder.

- Insert Centerlines, Center Marks and Annotations.

- Insert Custom Properties: Material, Description, DrawnBy, CompanyName, Number, Revision and DrawnDate.

- Material is 1060 Alloy.

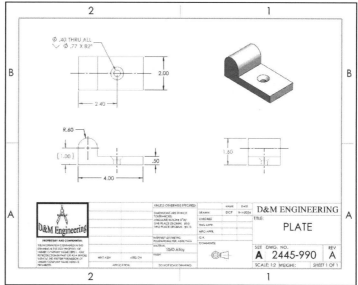

Exercise 5.7: LINKAGE-2 Drawing

- Create the LINKAGE-2 assembly drawing. Utilize the default A (ANSI) Landscape Sheet Format/Size. Open the LINKAGE-2 assembly from the Chapter 5 Homework\AirCylinder 2 folder.

- Insert an Isometric shaded view of the LINKAGE-2 Assembly.

- Define the PART NO. Property and the DESCRIPTION Property for the AXLE, FLATBAR-9HOLE, FLATBAR - 3HOLE and SHAFT COLLAR.

	Property Name	Type	Value / Text Expression	Evaluated Value		
1	SWFormatSize	Text	8.5in*11in	8.5in*11in		
2	Revision	Text	A	A		
3	DrawnBy	Text	DCP	DCP		
4	DrawnDate	Text	9-1-2024	9-1-2024		
5	Description	Text	LINKAGE-2	LINKAGE-2		
6	CompanyName	Text	D&M ENGINEERI	D&M ENGINEERING		
7	PartNo	Text	333-223-22	333-223-22		

- Save the LINKAGE-2 assembly to update the properties. Return to the LINKAGE-2 Drawing. Insert a Bill of Materials with Auto Balloons as illustrated.

- Insert the Company and Third Angle Projection icon.

- Insert Custom Properties: Description, DrawnBy, CompanyName, Number, Revision and DrawnDate.

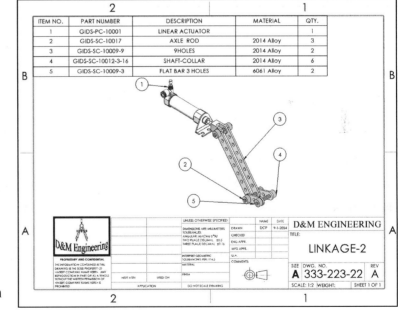

🔆 Use the Pack and Go option to save an assembly or drawing with references. The Pack and Go tool saves either to a folder or creates a zip file to e-mail. View SOLIDWORKS help for additional information.

Exercise 5.8: Vertical Section View Drawing

Create a Vertical Section view drawing. Modify the drawing view scale.

Create the A (ANSI) Landscape - IPS - Third Angle drawing as illustrated below. Do not display Tangent edges.

Insert Company and Third Angle projection icons. The icons are available in the homework folder.

Insert Custom Properties: Material, Description, DrawnBy, DrawnDate, CompanyName, etc.

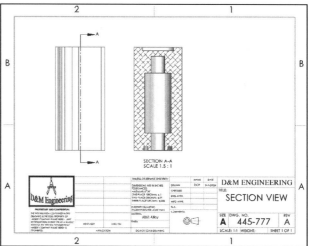

1. Open Section View from the Chapter 5 Homework\Section folder.

2. Create the drawing document.

3. Insert the Front view (Drawing View1).

4. Click inside the Drawing View1 view boundary. The Drawing View1 PropertyManager is displayed.

Display the origin on Sheet1.

5. Click View,Hide/Show, Origins from the Menu bar.

6. Click the Section View ⇆ drawing tool. The Section View PropertyManager is displayed.

7. Click the Section tab. Click the Vertical Cutting Line button.

8. Click the origin. Note: You can select the midpoint vs. the origin.

Place the Section view.

9. Click a position to the right of Drawing View1. The Section arrows point to the right. If required, click Flip direction. Check Auto hatching from the Section View box.

10. Check the Shaded With Edges option from the Display Style box. Click OK ✔ from the Section View A-A PropertyManager. Section View A-A is created and is displayed in the Drawing FeatureManager.

11. Click inside the Drawing View1 view boundary. The Drawing View1 PropertyManager is displayed.

12. Modify the Scale to 1.5:1. Click OK ✔ from the Drawing View1 PropertyManager. Both drawing views are modified (parent and child).

Appendix

SOLIDWORKS Keyboard Shortcuts

Below are some of the pre-defined keyboard shortcuts in SOLIDWORKS:

Action:	Key Combination:
Model Views	
Rotate the model horizontally or vertically	**Arrow** keys
Rotate the model horizontally or vertically 90 degrees	**Shift** + **Arrow** keys
Rotate the model clockwise or counterclockwise	**Alt** + left of right **Arrow** keys
Pan the model	**Ctrl** + **Arrow** keys
Magnifying glass	**g**
Zoom in	**Shift + z**
Zoom out	**z**
Zoom to fit	**f**
Previous view	**Ctrl + Shift + z**
View Orientation	
View Orientation menu	**Spacebar**
Front view	**Ctrl + 1**
Back view	**Ctrl + 2**
Left view	**Ctrl + 3**
Right view	**Ctrl + 4**
Top view	**Ctrl + 5**
Bottom view	**Ctrl + 6**
Isometric view	**Ctrl + 7**
NormalTo view	**Ctrl + 8**
Selection Filters	
Filter edges	**e**
Filter vertices	**v**
Filter faces	**x**
Toggle Selection Filter toolbar	**F5**
Toggle selection filters on/off	**F6**
File menu items	
New SOLIDWORKS document	**Ctrl + n**
Open document	**Ctrl + o**
Open From Web Folder	**Ctrl + w**
Make Drawing from Part	**Ctrl + d**
Make Assembly from Part	**Ctrl + a**
Save	**Ctrl +s**
Print	**Ctrl + p**
Additional items	
Access online help inside of PropertyManager or dialog box	**F1**
Rename an item in the FeatureManager design tree	**F2**

Action:	Key Combination:
Rebuild the model	**Ctrl + b**
Force rebuild - Rebuild the model and all its features	**Ctrl + q**
Redraw the screen	**Ctrl + r**
Cycle between open SOLIDWORKS document	**Ctrl + Tab**
Line to arc/arc to line in the Sketch	**a**
Undo	**Ctrl + z**
Redo	**Ctrl + y**
Cut	**Ctrl + x**
Copy	**Ctrl + c**
Paste	**Ctrl + v**
Delete	**Delete**
Next window	**Ctrl + F6**
Close window	**Ctrl + F4**
View previous tools	**s**
Selects all text inside an Annotations text box	**Ctrl + a**

In a sketch, the **Esc** key un-selects geometry items currently selected in the Properties box and Add Relations box.

In the model, the **Esc** key closes the PropertyManager and cancels the selections.

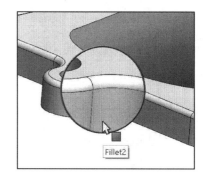

Use the **g** key to activate the Magnifying glass tool. Use the Magnifying glass tool to inspect a model and make selections without changing the overall view.

Use the **s** key to view/access previous command tools in the Graphics window.

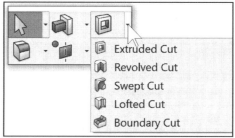

Modeling - Best Practices

Best practices are simply ways of bringing about better results in easier, more reliable ways. The Modeling - Best Practice list is a set of rules helpful for new users and users who are trying to experiment with the limits of the software.

These rules are not inflexible, but conservative starting places; they are concepts that you can default to, but that can be broken if you have good reason. The following is a list of suggested best practices:

- Create a folder structure (parts, drawings, assemblies, simulations, etc.). Organize into project or file folders.

- Construct sound document templates. The document template provides the foundation that all models are built on. This is especially important if working with other SOLIDWORKS users on the same project; it will ensure consistency across the project.

- Generate unique part filenames. SOLIDWORKS assemblies and drawings may pick up incorrect references if you use parts with identical names.

- Apply Custom Properties. Custom Properties is a great way to enter text-based information into the SOLIDWORKS parts. Users can view this information from outside the file by using applications such as Windows Explorer, SOLIDWORKS Explorer, and Product Data Management (PDM) applications.

- Understand part orientation. When you create a new part or assembly, the three default Planes (Front, Right and Top) are aligned with specific views. The plane you select for the Base sketch determines the orientation.

- Learn to sketch using automatic relations.

- Limit your usage of the Fixed constraint.

- Add geometric relations, then dimensions in a 2D sketch. This keeps the part from having too many unnecessary dimensions. This also helps to show the design intent of the model. Dimension what geometry you intend to modify or adjust.

- Fully define all sketches in the model. However, there are times when this is not practical, generally when using the Spline tool to create a freeform shape.

- When possible, make relations to sketches or stable reference geometry, such as the Origin or standard planes, instead of edges or faces. Sketches are far more stable than faces, edges, or model vertices, which change their internal ID at the slightest change and may disappear entirely with fillets, chamfers, split lines, and so on.

- Do not dimension to edges created by fillets or other cosmetic or temporary features.

- Apply names to sketches, features, dimensions, and mates that help to make their function clear.

- When possible, use feature fillets and feature patterns rather than sketch fillets and sketch patterns.

- Apply the Shell feature before the Fillet feature, and the inside corners remain perpendicular.

- Apply cosmetic fillets and chamfers last in the modeling procedure.

- Combine fillets into as few fillet features as possible. This enables you to control fillets that need to be controlled separately, such as fillets to be removed and simplified configurations.

- Create a simplified configuration when building very complex parts or working with large assemblies.

- Use symmetry during the modeling process. Utilize feature patterns and mirroring when possible. Think End Conditions.

- Use global variables and equations to control commonly applied dimensions (design intent).

- Add comments to equations to document your design intent. Place a single quote (') at the end of the equation, then enter the comment. Anything after the single quote is ignored when the equation is evaluated.

- Avoid redundant mates. Although SOLIDWORKS allows some redundant mates (all except distance and angle), these mates take longer to solve and make the mating scheme harder to understand and diagnose if problems occur.

- Fix modeling errors in the part or assembly when they occur. Errors cause rebuild time to increase, and if you wait until additional errors exist, troubleshooting will be more difficult.

- Create a Library of Standardized notes and parts.

- Utilize the Rollback bar. Troubleshoot feature and sketch errors from the top of the design tree.

- Determine the static and dynamic behavior of mates in each sub-assembly before creating the top-level assembly.

- Plan the assembly and sub-assemblies in an assembly layout diagram. Group components together to form smaller sub-assemblies.

- When you create an assembly document, the base component should be fixed, fully defined or mated to an axis about the assembly origin.

- In an assembly, group fasteners into a folder at the bottom of the FeatureManager. Suppress fasteners and their assembly patterns to save rebuild time and file size.

- When comparing mass, volume and other properties with assembly visualization, utilize similar units.

- Use limit mates sparingly because they take longer to solve and whenever possible, mate all components to one or two fixed components or references. Long chains of components take longer to solve and are more prone to mate errors.

Helpful On-line Information

The SOLIDWORKS URL:
http://www.SOLIDWORKS.com
contains information on Local
Resellers, Solution Partners,
Certifications, SOLIDWORKS users
groups and the **3D**EXPERIENCE
platform.

Use the SOLIDWORKS Task Pane
to obtain access to Customer Portals,
User Groups, Manufacturers,
Solution Partners, and the
3DEXPERIENCE Platform.

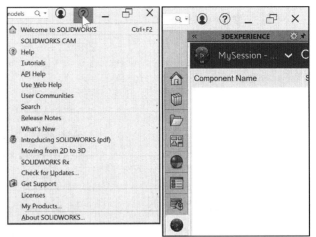

Helpful on-line SOLIDWORKS
information is available from the
following URLs:

- http://www.swugn.org/

List of all SOLIDWORKS User groups.

- https://www.solidworks.com/sw/edu
  cation/certification-programs-cad-
  students.htm

The SOLIDWORKS Academic
Certification Programs.

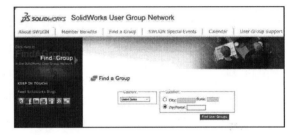

- http://www.solidworks.com/sw/in
  dustries/education/engineering-
  education-software.htm

The SOLIDWORKS Education
Program:

- visit
  https://3dexperience.virtualtester.c
  om/#home to obtain additional
  SOLIDWORKS Certification
  exam information.

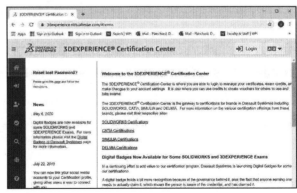

On-line tutorials are for educational
purposes only. Tutorials are
copyrighted by their respective
owners.

SOLIDWORKS Document Types

SOLIDWORKS has three main document file types: Part, Assembly and Drawing, but there are many additional supporting types that you may want to know. Below is a brief list of these supporting file types:

Design Documents	Description
.sldprt	SOLIDWORKS Part document
.slddrw	SOLIDWORKS Drawing document
.sldasm	SOLIDWORKS Assembly document

Templates and Formats	Description
.asmdot	Assembly Template
.asmprp	Assembly Template Custom Properties tab
.drwdot	Drawing Template
.drwprp	Drawing Template Custom Properties tab
.prtdot	Part Template
.prtprp	Part Template Custom Properties tab
.sldtbt	General Table Template
.slddrt	Drawing Sheet Template
.sldbombt	Bill of Materials Template (Table-based)
.sldholtbt	Hole Table Template
.sldrevbt	Revision Table Template
.sldwldbt	Weldment Cutlist Template
.xls	Bill of Materials Template (Excel-based)

Library Files	Description
.sldlfp	Library Part file
.sldblk	Blocks

Other	Description
.sldstd	Drafting standard
.sldmat	Material Database
.sldclr	Color Palette File
.xls	Sheet metal gauge table

GLOSSARY

Alphabet of Lines: Each line on a technical drawing has a definite meaning and is drawn in a certain way. The line conventions recommended by the American National Standards Institute (ANSI) are presented in this text.

Alternate Position View: A drawing view superimposed in phantom lines on the original view. Utilized to show range of motion of an assembly.

Anchor Point: The origin of the Bill of Material in a sheet format.

Annotation: An annotation is a text note or a symbol that adds specific information and design intent to a part, assembly, or drawing. Annotations in a drawing include specific note, hole callout, surface finish symbol, datum feature symbol, datum target, geometric tolerance symbol, weld symbol, balloon and stacked balloon, center mark, centerline marks, area hatch and block.

ANSI: American National Standards Institute.

Area Hatch: Apply a crosshatch pattern or solid fill to a model face, to a closed sketch profile, or to a region bounded by a combination of model edges and sketch entities. Area hatch can be applied only in drawings.

ASME: American Society of Mechanical Engineering, publisher of ASME Y14 Engineering Drawing and Documentation Practices that controls drawing, dimensioning and tolerancing.

Assembly: An assembly is a document in which parts, features and other assemblies (sub-assemblies) are put together. A part in an assembly is called a component. Adding a component to an assembly creates a link between the assembly and the component. When SOLIDWORKS opens the assembly, it finds the component file to show it in the assembly. Changes in the component are automatically reflected in the assembly. The filename extension for a SOLIDWORKS assembly file name is *.sldasm.

Attachment Point: An attachment point is the end of a leader that attaches to an edge, vertex, or face in a drawing sheet.

AutoDimension: The Autodimension tool provides the ability to insert reference dimensions into drawing views such as baseline, chain, and ordinate dimensions.

Auxiliary View: An Auxiliary View is similar to a Projected View, but it is unfolded normal to a reference edge in an existing view.

AWS: American Welding Society, publisher of AWS A2.4, Standard Location of Elements of a Welding Symbol.

Axonometric Projection: A type of parallel projection, more specifically a type of orthographic projection, used to create a pictorial drawing of an object, where the object is rotated along one or more of its axes relative to the plane of projection.

Balloon: A balloon labels the parts in the assembly and relates them to item numbers on the bill of materials (BOM) added in the drawing. The balloon item number corresponds to the order in the Feature Tree. The order controls the initial BOM Item Number.

Baseline Dimensions: Dimensions referenced from the same edge or vertex in a drawing view.

Bill of Materials: A table inserted into a drawing to keep a record of the parts and materials used in an assembly.

Block: A symbol in the drawing that combines geometry into a single entity.

BOM: Abbreviation for Bill of Materials.

Broken-out Section: A broken-out section exposes inner details of a drawing view by removing material from a closed profile. In an assembly, the Broken-out Section displays multiple components.

CAD: The use of computer technology for the design of objects, real or virtual. CAD often involves more than just shapes.

Cartesian Coordinate System: Specifies each point uniquely in a plane by a pair of numerical coordinates, which are the signed distances from the point to two fixed perpendicular directed lines, measured in the same unit of length. Each reference line is called a coordinate axis or just axis of the system, and the point where they meet is its origin.

Cell: Area to enter a value in an EXCEL spreadsheet, identified by a Row and Column.

Center Mark: A cross that marks the center of a circle or arc.

Centerline: An axis of symmetry in a sketch or drawing displayed in a phantom font.

CommandManager: The CommandManager is a Context-sensitive toolbar that dynamically updates based on the toolbar you want to access. By default, it has toolbars embedded in it based on the document type. When you click a tab below the Command Manager, it updates to display that toolbar. For example, if you click the Sketch tab, the Sketch toolbar is displayed.

Component: A part or sub-assembly within an assembly.

ConfigurationManager: The ConfigurationManager is located on the left side of the SOLIDWORKS window and provides the means to create, select and view multiple configurations of parts and assemblies in an active document. You can split the

ConfigurationManager and either display two ConfigurationManager instances, or combine the ConfigurationManager with the FeatureManager design tree, PropertyManager or third-party applications that use the panel.

Configurations: Variations of a part or assembly that control dimensions, display and state of a model.

Coordinate System: SOLIDWORKS uses a coordinate system with origins. A part document contains an original origin. Whenever you select a plane or face and open a sketch, an origin is created in alignment with the plane or face. An origin can be used as an anchor for the sketch entities, and it helps orient perspective of the axes. A three-dimensional reference triad orients you to the X, Y, and Z directions in part and assembly documents.

Copy and Paste: Utilize copy/paste to copy views from one sheet to another sheet in a drawing or between different drawings.

Cosmetic Thread: An annotation that represents threads.

Crosshatch: A pattern (or fill) applied to drawing views such as section views and broken-out sections.

Cursor Feedback: The system feedback symbol indicates what you are selecting or what the system is expecting you to select. As you move the mouse pointer across your model, system feedback is provided.

Datum Feature: An annotation that represents the primary, secondary and other reference planes of a model utilized in manufacturing.

Depth: The horizontal (front to back) distance between two features in frontal planes. Depth is often identified in the shop as the thickness of a part or feature.

Design Table: An Excel spreadsheet that is used to create multiple configurations in a part or assembly document.

Detail View: A portion of a larger view, usually at a larger scale than the original view. Create a detail view in a drawing to display a portion of a view, usually at an enlarged scale. This detail may be of an orthographic view, a non-planar (isometric) view, a section view, a crop view, an exploded assembly view or another detail view.

Detailing: Detailing refers to the SOLIDWORKS module used to insert, add and modify dimensions and notes in an engineering drawing.

Dimension Line: A line that references dimension text to extension lines indicating the feature being measured.

Dimension Tolerance: Controls the dimension tolerance values and the display of non-integer dimensions. The tolerance types are *None, Basic, Bilateral, Limit, Symmetric, MIN, MAX, Fit, Fit with tolerance* or *Fit (tolerance only)*.

Dimension: A value indicating the size of the 2D sketch entity or 3D feature. Dimensions in a SOLIDWORKS drawing are associated with the model, and changes in the model are reflected in the drawing, if you DO NOT USE DimXpert.

Dimensioning Standard - Metric: - ASME standards for the use of metric dimensioning required all the dimensions to be expressed in millimeters (mm). The (mm) is not needed on each dimension, but it is used when a dimension is used in a notation. No trailing zeroes are used. The Metric or International System of Units (S.I.) unit system in drafting is also known as the Millimeter, Gram Second (MMGS) unit system.

Dimensioning Standard - U.S: - ASME standard for U.S. dimensioning uses the decimal inch value. When the decimal inch system is used, a zero is not used to the left of the decimal point for values less than one inch, and trailing zeroes are used. The U.S. unit system is also known as the Inch, Pound, Second (IPS) unit system.

DimXpert for Parts: A set of tools that applies dimensions and tolerances to parts according to the requirements of the ASME Y.14.41-2009 standard.

DimXpertManager: The DimXpertManager lists the tolerance features defined by DimXpert for a part. It also displays DimXpert tools that you use to insert dimensions and tolerances into a part. You can import these dimensions and tolerances into drawings. DimXpert is not associative.

Document: In SOLIDWORKS, each part, assembly, and drawing is referred to as a document, and each document is displayed in a separate window.

Drawing Sheet: A page in a drawing document.

Drawing Template: A document that is the foundation of a new drawing. The drawing template contains document properties and user-defined parameters such as sheet format. The extension for the drawing template filename is .DRWDOT.

Drawing: A 2D representation of a 3D part or assembly. The extension for a SOLIDWORKS drawing file name is .SLDDRW. Drawing refers to the SOLIDWORKS module used to insert, add, and modify views in an engineering drawing.

Edit Sheet Format: The drawing sheet contains two modes. Utilize the Edit Sheet Format command to add or modify notes and Title block information. Edit in the Edit Sheet Format mode.

Edit Sheet: The drawing sheet contains two modes. Utilize the Edit Sheet command to insert views and dimensions.

eDrawing: A compressed document that does not require the referenced part or assembly. eDrawings are animated to display multiple views in a drawing.

Empty View: An Empty View creates a blank view not tied to a part or assembly document.

Engineering Graphics: Translates ideas from design layouts, specifications, rough sketches, and calculations of engineers & architects into working drawings, maps, plans and illustrations which are used in making products.

Equation: Creates a mathematical relation between sketch dimensions, using dimension names as variables, or between feature parameters, such as the depth of an extruded feature or the instance count in a pattern.

Exploded view: A configuration in an assembly that displays its components separated from one another.

Export: The process to save a SOLIDWORKS document in another format for use in other CAD/CAM, rapid prototyping, web or graphics software applications.

Extension Line: The line extending from the profile line indicating the point from which a dimension is measured.

Extruded Cut Feature: Projects a sketch perpendicular to a Sketch plane to remove material from a part.

Face: A selectable area (planar or otherwise) of a model or surface with boundaries that help define the shape of the model or surface. For example, a rectangular solid has six faces.

Family Cell: A named empty cell in a Design Table that indicates the start of the evaluated parameters and configuration names. Locate Comments in a Design Table to the left or above the Family Cell.

Fasteners: Includes Bolts and nuts (threaded), Set screws (threaded), Washers, Keys, and Pins to name a few. Fasteners are not a permanent means of assembly such as welding or adhesives.

Feature: Features are geometry building blocks. Features add or remove material. Features are created from 2D or 3D sketched profiles or from edges and faces of existing geometry.

FeatureManager: The FeatureManager design tree located on the left side of the SOLIDWORKS window provides an outline view of the active part, assembly, or drawing. This makes it easy to see how the model or assembly was constructed or to examine the various sheets and views in a drawing. The FeatureManager and the Graphics window are dynamically linked. You can select features, sketches, drawing views and construction geometry in either pane.

First Angle Projection: In First Angle Projection the Top view is looking at the bottom of the part. First Angle Projection is used in Europe and most of the world. However, America and Australia use a method known as Third Angle Projection.

Fully defined: A sketch where all lines and curves in the sketch, and their positions, are described by dimensions or relations, or both, and cannot be moved. Fully defined sketch entities are shown in black.

Foreshortened radius: Helpful when the centerpoint of a radius is outside of the drawing or interferes with another drawing view: Broken Leader.

Foreshortening: The way things appear to get smaller in both height and depth as they recede into the distance.

French curve: A template made out of plastic, metal or wood composed of many different curves. It is used in manual drafting to draw smooth curves of varying radii.

Fully Defined: A sketch where all lines and curves in the sketch, and their positions, are described by dimensions or relations, or both, and cannot be moved. Fully defined sketch entities are displayed in black.

Geometric Tolerance: A set of standard symbols that specify the geometric characteristics and dimensional requirements of a feature.

Glass Box method: A traditional method of placing an object in an *imaginary glass box* to view the six principle views.

Global Coordinate System: Directional input refers by default to the Global coordinate system (X-, Y- and Z-), which is based on Plane1 with its origin located at the origin of the part or assembly.

Graphics Window: The area in the SOLIDWORKS window where the part, assembly, or drawing is displayed.

Grid: A system of fixed horizontal and vertical divisions.

Handle: An arrow, square or circle that you drag to adjust the size or position of an entity such as a view or dimension.

Heads-up View Toolbar: A transparent toolbar located at the top of the Graphic window.

Height: The vertical distance between two or more lines or surfaces (features) which are in horizontal planes.

Hidden Lines Removed (HLR): A view mode. All edges of the model that are not visible from the current view angle are removed from the display.

Hidden Lines Visible (HLV): A view mode. All edges of the model that are not visible from the current view angle are shown gray or dashed.

Hole Callouts: Hole callouts are available in drawings. If you modify a hole dimension in the model, the callout updates automatically in the drawing if you did not use DimXpert.

Hole Table: A table in a drawing document that displays the positions of selected holes from a specified origin datum. The tool labels each hole with a tag. The tag corresponds to a row in the table.

Import: The ability to open files from other software applications into a SOLIDWORKS document. The A-size sheet format was created as an AutoCAD file and imported into SOLIDWORKS.

Isometric Projection: A form of graphical projection, more specifically, a form of axonometric projection. It is a method of visually representing three-dimensional objects in two dimensions, in which the three coordinate axes appear equally foreshortened and the angles between any two of them are 120º.

Layers: Simplifies a drawing by combining dimensions, annotations, geometry and components. Properties such as display, line style and thickness are assigned to a named layer.

Leader: A solid line created from an annotation to the referenced feature.

Line Format: A series of tools that controls Line Thickness, Line Style, Color, Layer and other properties.

Local (Reference) Coordinate System: Coordinate system other than the Global coordinate system. You can specify restraints and loads in any desired direction.

Lock Sheet Focus: Adds sketch entities and annotations to the selected sheet. Double-click the sheet to activate Lock Sheet Focus. To unlock a sheet, right-click and select Unlock Sheet Focus or double click inside the sheet boundary.

Lock View Position: Secures the view at its current position in the sheet. Right-click in the drawing view to Lock View Position. To unlock a view position, right-click and select Unlock View Position.

Mass Properties: The physical properties of a model based upon geometry and material.

Menus: Menus provide access to the commands that the SOLIDWORKS software offers. Menus are Context-sensitive and can be customized through a dialog box.

Model Item: Provides the ability to insert dimensions, annotations, and reference geometry from a model document (part or assembly) into a drawing.

Model View: A specific view of a part or assembly. Standard named views are listed in the view orientation dialog box such as isometric or front. Named views can be user-defined names for a specific view.

Model: 3D solid geometry in a part or assembly document. If a part or assembly document contains multiple configurations, each configuration is a separate model.

Motion Studies: Graphical simulations of motion and visual properties with assembly models. Analogous to a configuration, they do not actually change the original assembly model or its properties. They display the model as it changes based on simulation elements you add.

Mouse Buttons: The left, middle, and right mouse buttons have distinct meanings in SOLIDWORKS. Use the middle mouse button to rotate and Zoom in/out on the part or assembly document.

Oblique Projection: A simple type of graphical projection used for producing pictorial, two-dimensional images of three-dimensional objects.

OLE (Object Linking and Embedding): A Windows file format. A company logo or EXCEL spreadsheet placed inside a SOLIDWORKS document are examples of OLE files.

Ordinate Dimensions: Chain of dimensions referenced from a zero ordinate in a drawing or sketch.

Origin: The model origin is displayed in blue and represents the (0,0,0) coordinate of the model. When a sketch is active, a sketch origin is displayed in red and represents the (0,0,0) coordinate of the sketch. Dimensions and relations can be added to the model origin but not to a sketch origin.

Orthographic Projection: A means of representing a three-dimensional object in two dimensions. It is a form of parallel projection, where the view direction is orthogonal to the projection plane, resulting in every plane of the scene appearing in affine transformation on the viewing surface.

Parametric Note: A Note annotation that links text to a feature dimension or property value.

Parent View: A Parent view is an existing view on which other views are dependent.

Part Dimension: Used in creating a part, they are sometimes called construction dimensions.

Part: A 3D object that consists of one or more features. A part inserted into an assembly is called a component. Insert part views, feature dimensions and annotations into 2D drawing. The extension for a SOLIDWORKS part filename is .SLDPRT.

Perspective Projection: The two most characteristic features of perspective are that objects are drawn: smaller as their distance from the observer increases and foreshortened: the size of an object's dimensions along the line of sight are relatively shorter than dimensions across the line of sight.

Plane: To create a sketch, choose a plane. Planes are flat and infinite. Planes are represented on the screen with visible edges.

Precedence of Line Types: When obtaining orthographic views, it is common for one type of line to overlap another type. When this occurs, drawing conventions have established an order of precedence.

Precision: Controls the number of decimal places displayed in a dimension.

Projected View: Projected views are created for Orthogonal views using one of the following tools: Standard 3 View, Model View or the Projected View tool from the View Layout toolbar.

Properties: Variables shared between documents through linked notes.

PropertyManager: Most sketch, feature, and drawing tools in SOLIDWORKS open a PropertyManager located on the left side of the SOLIDWORKS window. The PropertyManager displays the properties of the entity or feature so you specify the properties without a dialog box covering the Graphics window.

RealView: Provides a simplified way to display models in a photo-realistic setting using a library of appearances and scenes. RealView requires graphics card support and is memory intensive.

Rebuild: A tool that updates (or regenerates) the document with any changes made since the last time the model was rebuilt. Rebuild is typically used after changing a model dimension.

Reference Dimension: Dimensions added to a drawing document are called Reference dimensions and are driven; you cannot edit the value of reference dimensions to modify the model. However, the values of reference dimensions change when the model dimensions change.

Relation: A relation is a geometric constraint between sketch entities or between a sketch entity and a plane, axis, edge or vertex.

Relative view: The Relative View defines an Orthographic view based on two orthogonal faces or places in the model.

Revision Table: The Revision Table lists the Engineering Change Orders (ECO), in a table form, issued over the life of the model and the drawing. The current Revision letter or number is placed in the Title block of the Drawing.

Right-Hand Rule: Is a common mnemonic for understanding notation conventions for vectors in 3 dimensions.

Rollback: Suppresses all items below the rollback bar.

Scale: A relative term meaning "size" in relationship to some system of measurement.

Section Line: A line or centerline sketched in a drawing view to create a section view.

Section Scope: Specifies the components to be left uncut when you create an assembly drawing section view.

Section View: You create a section view in a drawing by cutting the parent view with a cutting, or section line. The section view can be a straight cut section or an offset section defined by a stepped section line. The section line can also include concentric arcs. Create a Section View in a drawing by cutting the Parent view with a section line.

Sheet Format: A document that contains the following: page size and orientation, standard text, borders, logos, and Title block information. Customize the Sheet format to save time. The extension for the Sheet format filename is .SLDDRT.

Sheet Properties: Sheet Properties display properties of the selected sheet. Sheet Properties define the following: Name of the Sheet, Sheet Scale, Type of Projection (First angle or Third angle), Sheet Format, Sheet Size, View label, and Datum label.

Sheet: A page in a drawing document.

Silhouette Edge: A curve representing the extent of a cylindrical or curved face when viewed from the side.

Sketch: The name to describe a 2D profile is called a sketch. 2D sketches are created on flat faces and planes within the model. Typical geometry types are lines, arcs, corner rectangles, circles, polygons, and ellipses.

Spline: A sketched 2D or 3D curve defined by a set of control points.

Stacked Balloon: A group of balloons with only one leader. The balloons can be stacked vertically (up or down) or horizontally (left or right).

Standard views: The three orthographic projection views, Front, Top and Right positioned on the drawing according to First angle or Third angle projection.

Suppress: Removes an entity from the display and from any calculations in which it is involved. You can suppress features, assembly components, and so on. Suppressing an entity does not delete the entity; you can unsuppress the entity to restore it.

Surface Finish: An annotation that represents the texture of a part.

System Feedback: Feedback is provided by a symbol attached to the cursor arrow indicating your selection. As the cursor floats across the model, feedback is provided in the form of symbols riding next to the cursor.

System Options: System Options are stored in the registry of the computer. System Options are not part of the document. Changes to the System Options affect all current and future documents. There are hundreds of Systems Options.

Tangent Edge: The transition edge between rounded or filleted faces in hidden lines visible or hidden lines removed modes in drawings.

Task Pane: The Task Pane is displayed when you open the SOLIDWORKS software. It contains the following tabs: SOLIDWORKS Resources, Design Library, File Explorer, Search, View Palette, Document Recovery and RealView/PhotoWorks.

Templates: Templates are part, drawing and assembly documents that include user-defined parameters and are the basis for new documents.

Third Angle Projection: In Third angle projection the Top View is looking at the Top of the part. First Angle Projection is used in Europe and most of the world. America and Australia use the Third Angle Projection method.

Thread Class or Fit: Classes of fit are tolerance standards; they set a plus or minus figure that is applied to the pitch diameter of bolts or nuts. The classes of fit used with almost all bolts sized in inches are specified by the ANSI/ASME Unified Screw Thread standards (which differ from the previous American National standards).

Thread Lead: The distance advanced parallel to the axis when the screw is turned one revolution. For a single thread, lead is equal to the pitch; for a double thread, lead is twice the pitch.

Tolerance: The permissible range of variation in a dimension of an object. Tolerance may be specified as a factor or percentage of the nominal value, a maximum deviation from a nominal value, an explicit range of allowed values, be specified by a note or published standard with this information, or be implied by the numeric accuracy of the nominal value.

Toolbars: The toolbar menus provide shortcuts enabling you to access the most frequently used commands. Toolbars are Context-sensitive and can be customized through a dialog box.

T-Square: A technical drawing instrument, primarily a guide for drawing horizontal lines on a drafting table. It is used to guide the triangle that draws vertical lines. Its name comes from the general shape of the instrument where the horizontal member of the T slides on the side of the drafting table. Common lengths are 18", 24", 30", 36" and 42".

Under-defined: A sketch is under defined when there are not enough dimensions and relations to prevent entities from moving or changing size.

Units: Used in the measurement of physical quantities. Decimal inch dimensioning and Millimeter dimensioning are the two types of common units specified for engineering parts and drawings.

Vertex: A point at which two or more lines or edges intersect. Vertices can be selected for sketching, dimensioning, and many other operations.

View Palette: Use the View Palette, located in the Task Pane, to insert drawing views. It contains images of standard views, annotation views, section views, and flat patterns (sheet metal parts) of the selected model. You can drag views onto the drawing sheet to create a drawing view.

Weld Bead: An assembly feature that represents a weld between multiple parts.

Weld Finish: A weld symbol representing the parameters you specify.

Weld Symbol: An annotation in the part or drawing that represents the parameters of the weld.

Width: The horizontal distance between surfaces in profile planes. In the machine shop, the terms length and width are used interchangeably.

Zebra Stripes: Simulate the reflection of long strips of light on a very shiny surface. They allow you to see small changes in a surface that may be hard to see with a standard display.

INDEX